God Made
ANIMALS

WORKBOOK

Copyright © 2019 by Generations. All rights reserved. No part of this book may be reproduced in any form or by any means without permission in writing from the publisher.

1st Printing, 2019.

Printed in the United States of America.

ISBN: 978-1-7332304-2-1

Cover Design: Justin Turley
Interior Design: Justin, Turley, Rei Suzuki

Published by:
Generations
19039 Plaza Drive Ste 210
Parker, Colorado 80134
Generations.org

Unless otherwise noted, Scripture taken from the New King James Version®. Copyright © 1982 by Thomas Nelson. Used by permission. All rights reserved.

For more information on this and other titles from Generations,
visit Generations.org or call (888) 389-9080.

TABLE OF CONTENTS

Introduction · 5
Lesson Schedule · 12

Chapter 1— Science from a Biblical Worldview · · · · · · · · · · · · · · 23
Chapter 2—God's Amazing Creative Power · · · · · · · · · · · · · · · · · 31
Chapter 3—God Made Animals · 39
Chapter 4—Animals Without Backbones · · · · · · · · · · · · · · · · · · 47
Chapter 5—How Insects Serve God and His Creation · · · · · · · · · 57
Chapter 6—God Made Birds · 65
Chapter 7—God's Awe-Inspiring Designs for Bird Function · · · · 73
Chapter 8—God's Command to Rule Over the Birds · · · · · · · · · · 81
Chapter 9—God Made Fish · 87
Chapter 10—God Designed Fish for a Purpose · · · · · · · · · · · · · · 95
Chapter 11—Taking Dominion of the Fish · · · · · · · · · · · · · · · · · 101
Chapter 12—God Made Amphibians · 107
Chapter 13—God Made Reptiles · 117
Chapter 14—God Made Both Great and Small Reptiles · · · · · · · 125
Chapter 15—God Made Mammals · 133
Chapter 16—Taking Dominion of the Mammal World · · · · · · · · 143
Final Examination · 153

Answer Key · 158

INTRODUCTION

This course will represent a paradigm shift in the way that the student understands science. A biblical worldview perspective of science always puts God at the center. The purpose of science is to enjoy God in the context of His creation and praise Him for His marvelous works.

Thus, this course will provide devotional reading in many Scriptures throughout, as we introduce the animal creation to the students. There will be many opportunities provided for prayer and singing of praise. These elements are core to a Christian view of science. As the teacher/parent disciples the child in the study of science, we hope the student develops a Godward view of science and all of life.

If education will be truly effective, the student must be well aware of the purpose of the study throughout. Hardly a page of the text should go by without the student realizing the high significance and purpose of the study. For the Christian the purpose of science is absolutely clear. It is for praise and dominion. We glory in God's nature and His works, and we seek to fulfill our role in ruling over the natural world (as the Lord commanded man to do at the beginning).

If there is a clear integration of praise and life application on every page, the material will be much better retained. The perpetual spewing of disconnected and purposeless facts in education does little for retention. When science is given meaning and purpose on every page, the student will be much more likely retain the material and apply it in a meaningful way in his life.

There must be an integration of the material the student is learning into real life. Great efforts have been taken to provide suggestions how this might be done. Laboratory work and experiments are interesting and helpful, but they are not enough. The student must find ways to take the scientific knowledge he has picked up and apply it to his own life, to his own home and garden, and to his own community.

For student in 5-7 grade, Generations introduces a biblical worldview into field of science in the most winsome way possible. Captured in this introduction to animal biology are the most amazing facts and the most interesting facets of God's creation. The course is filled with Scripture and with a sense of wonder and praise, which is only fitting for a Christian child's education. Prayers of praise and hymns will play an important part in the course. Critical elements of the biblical worldview in the scientific field are laid out, to counter the false view of science that has become increasingly destructive in the present day.

Included in this course are the world of vertebrates, invertebrates, warm-blooded animals, cold-blooded animals, insects, the bombardier beetle, the butterfly, the spider, the mosquito, the dung beetle, the house fly, the silkworm, grasshoppers, the honeybee, the leech, the octopus, the jellyfish, the penguin, the chicken, the crow, the pigeon, as well as the wonderful world of fish and mammals.

Also covered in this course is basic genetics, basic taxonomy, the basics of bird flight, animal domestication, and real life science application opportunities for the student. The student is introduced to man's responsibility to take dominion over insects, birds, fish, and domesticated animals, as one of the most critical elements of science (from a biblical perspective).

THIS STUDENT WORKBOOK INCLUDES:

1. Lesson Schedule
2. Study Questions
3. Scripture Exercises
4. Spiritual Life Application Questions
5. Special Observation and Dissections Projects
6. Final Examination
8. Answer Key

COURSE OBJECTIVES

This course is dedicated to the glory of God and to the preeminence of the Lord Jesus Christ in all things. The essential objectives for the course must therefore be:

1. That all who study this course would give God the glory for His sovereignty, His power, His

INTRODUCTION

goodness, His wisdom, His judgments, and His mercy.

2. That our children would realize that Jesus Christ is the Creator of all things and by Him all things consist.

3. That our children would recognize that Jesus is right now ruling as King of kings and Lord of lords.

4. That our children would immediately realize the purpose for science on every page and in every lesson — to glorify God and to wisely and obediently rule over the animal creation.

5. That our children will learn to life integrate the knowledge they obtain of God's world.

6. That our children would know Scripture better and see its amazing relevance to every part of life (including science), especially as the many Scripture references are read and meditated upon throughout the duration of this course.

7. That our children would learn to be more grateful and more ready to give God the praise and the glory for His goodness to us.

TEACHING METHOD

In order for this course to have maximum effect, the teacher/parent should:

1. Realize the joy and excitement of exploring God's world
2. Love God
3. Seek to learn more of the awesomeness of God manifested in His creative work, and share that enthusiasm with the children
4. Accept the obligation to follow through on knowledge by real life application

This curriculum and lesson schedule is laid out in a carefully designed manner, such that the lesson culminates in praise and practical life application. The following presents the order of the learning process:

1. Read the material.
2. Pray, sing, and worship God.
3. Watch excellently-produced videos to better understand the material.
4. Answer study questions and review Bible passages.
5. Make spiritual life application.
6. Observe through dissections and other projects.
7. Take dominion, using the "Do" projects contained in the textbook.

Almost every chapter in the text includes real-life application exercises and Observation/Dissection exercises in the workbook. Some of these exercises will be more time-intensive than others. However, there are usually some exercises that require far less time to complete. Select the more time-intensive exercise carefully, and we would only recommend 1-2 time-intensive exercises for each of the "Do" projects and the Observation/Dissection categories of exercises.

The parent/teacher may consider reading the material out loud. The text is designed to be engaging to children of all ages as well as adults.

LESSON SCHEDULE

The lesson schedule is provided as a suggestion—teachers/parents and students may adapt the schedule to suit their needs. The lesson schedule is based on a 36-week school year divided into two semesters.

COMPLETING CHAPTER ASSIGNMENTS

While reading the text, the student should carefully consider all of the Scriptures provided. The Scriptures provide the most essential elements of knowledge by which we understand God's world.

The key terms and animal designations are emboldened and colored in orange. The students should pay close attention to these as they will be referred to in the study questions and exam. Upon completion of reading, students may complete the chapter assignments open-book.

COMPLETING THE FINAL EXAMINATION

The final examination must be completed closed-book and the student may not use notes, workbook, or text.

INTRODUCTION

GRADING CHAPTER ASSIGNMENTS

The Teacher/Parent may determine for themselves how they would grade the assignments. The following is recommended:

The Study Questions and the Scripture Exercises are best suited for grading. Provide one point for each of these questions and exercises. For each chapter assignment, divide total number of questions answered correctly with the total number of questions possible to calculate the percentage.

For example, if 8 out of 11 questions were correct, then the percentage grade for that assignment will be 72%.

$$8 / 11 = 72\%$$

If the student receives less than 90% correct answers, it is highly recommended that he/she re-read the chapter and make corrections for the questions missed.

GRADING PROJECTS

It is recommended that the projects be graded on the basis of completion or participation. If the student completed the project, he gets 100% for that project. If he left the project only half complete, he gets 0% for that project. If the student completes 14 out 16 of the projects assigned, the student's score is 14/16 = 87.5%

FINAL COURSE GRADE VALUES

To calculate the final course grade the parent/teacher may use the following recommended weighted score:

Completion of projects - 40%
Study Questions/Scripture Exercises - 40%
Final Examination - 20%

Final Grade = (Project percentage x 0.40) + (Study Question percentage x 0.40) + Final Exam percentage x 0.20)

The following may be used for grade values when grading chapter assignments, exams, or projects:

90 to 100 percent = A
80-89 percent = B
70 to 79 percent = C
60 to 69 percent = D
0 to 59 percent = F

The Generations Curriculum Team
July 2019, AD

GOD MADE ANIMALS - WORKBOOK

Lesson Schedule

Date	Day	Assignment	Due Date	✔	Grade
First Semester–First Quarter					
Week 1	1	Read **Introduction** and **Chapter 1** Pray and Sing Hymn			
	2				
	3	Watch **Chapter 1** Videos			
	4	Complete **Chapter 1** Study Questions, Scripture, and Spiritual Life Application			
	5				
Week 2	1				
	2				
	3	Complete **Chapter 1** Observation or Dissection			
	4				
	5				
Week 3	1	Read **Chapter 2** Pray and Sing Hymn			
	2				
	3	Watch **Chapter 2** Videos			
	4	Complete **Chapter 2** Study Questions, Scripture, and Spiritual Life Application			
	5				
Week 4	1				
	2				
	3	Complete **Chapter 2** Observation or Dissection			
	4				
	5				

LESSON SCHEDULE

Date	Day	Assignment	Due Date	✔	Grade
Week 5	1	Read **Chapter 3** Pray and Sing Hymn			
	2				
	3	Watch **Chapter 3** Videos			
	4	Complete **Chapter 3** Study Questions, Scripture, and Spiritual Life Application			
	5				
Week 6	1	Complete **Chapter 3** Observation or Dissection			
	2				
	3				
	4				
	5				
Week 7	1	Read **Chapter 4** Pray and Sing Hymn			
	2				
	3	Watch **Chapter 4** Videos			
	4	Complete **Chapter 4** Study Questions, Scripture, and Spiritual Life Application			
	5				
Week 8	1	Complete **Chapter 4** "Do" Activities and Observation or Dissection			
	2				
	3				
	4				
	5				

Date	Day	Assignment	Due Date	✔	Grade
Week 9	1	Flex Week			
	2				
	3				
	4				
	5				
First Semester–Second Quarter					
Week 1	1	Read **Chapter 5** Pray and Sing Hymn			
	2				
	3	Watch **Chapter 5** Videos			
	4	Complete **Chapter 5** Study Questions, Scripture, and Spiritual Life Application			
	5				
Week 2	1	Complete **Chapter 5** "Do" Activities and Observation or Dissection			
	2				
	3				
	4				
	5				
Week 3	1	Read **Chapter 6** Pray and Sing Hymn			
	2				
	3	Watch **Chapter 6** Videos			
	4	Complete **Chapter 6** Study Questions, Scripture, and Spiritual Life Application			
	5				

LESSON SCHEDULE

Date	Day	Assignment	Due Date	✔	Grade
Week 4	1	Complete **Chapter 6** "Do" Activities and Observation or Dissection			
	2				
	3				
	4				
	5				
Week 5	1	Read **Chapter 7** Pray and Sing Hymn			
	2				
	3	Watch **Chapter 7** Videos			
	4	Complete **Chapter 7** Study Questions, Scripture, and Spiritual Life Application			
	5				
Week 6	1	Complete **Chapter 7** Observation or Dissection			
	2				
	3				
	4				
	5				
Week 7	1	Read **Chapter 8** Pray and Sing Hymn			
	2				
	3	Watch **Chapter 8** Videos			
	4	Complete **Chapter 8** Study Questions, Scripture, and Spiritual Life Application			
	5				

GOD MADE ANIMALS - WORKBOOK

Date	Day	Assignment	Due Date	✔	Grade
Week 8	1	Complete **Chapter 8** "Do" Activities and Observation or Dissection			
	2				
	3				
	4				
	5				
Week 9	1	Read **Chapter 9** Pray and Sing Hymn			
	2				
	3	Watch **Chapter 9** Videos			
	4	Complete **Chapter 9** Study Questions, Scripture, and Spiritual Life Application			
	5				

Second Semester–Third Quarter

Date	Day	Assignment	Due Date	✔	Grade
Week 1	1	Complete **Chapter 9** Observation or Dissection			
	2				
	3				
	4				
	5				
Week 2	1	Read **Chapter 10** Pray and Sing Hymn			
	2				
	3	Watch **Chapter 10** Videos			
	4	Complete **Chapter 10** Study Questions, Scripture, and Spiritual Life Application			
	5				

LESSON SCHEDULE

Date	Day	Assignment	Due Date	✔	Grade
Week 3	1	Complete **Chapter 10** Observation or Dissection			
	2				
	3				
	4				
	5				
Week 4	1	Read **Chapter 11** Pray and Sing Hymn			
	2				
	3	Watch **Chapter 11** Videos			
	4	Complete **Chapter 11** Study Questions, Scripture, and Spiritual Life Application			
	5				
Week 5	1	Complete **Chapter 11** "Do" Activities and Observation or Dissection			
	2				
	3				
	4				
	5				
Week 6	1	Read **Chapter 12** Pray and Sing Hymn			
	2				
	3	Watch **Chapter 12** Videos			
	4	Complete **Chapter 12** Study Questions, Scripture, and Spiritual Life Application			
	5				

GOD MADE ANIMALS - WORKBOOK

Date	Day	Assignment	Due Date	✔	Grade
Week 7	1	Complete **Chapter 12** "Do" Activities and Observation or Dissection			
	2				
	3				
	4				
	5				
Week 8	1	Flex Week			
	2				
	3				
	4				
	5				
Week 9	1	Read **Chapter 13** Pray and Sing Hymn			
	2				
	3	Watch **Chapter 13** Videos			
	4	Complete **Chapter 13** Study Questions, Scripture, and Spiritual Life Application			
	5				
		Second Semester–Fourth Quarter			
Week 1	1	Complete **Chapter 13** "Do" Activities and Research, Observation, and Dissection			
	2				
	3				
	4				
	5				

LESSON SCHEDULE

Date	Day	Assignment	Due Date	✔	Grade
Week 2	1	Read **Chapter 14** Pray and Sing Hymn			
	2				
	3	Watch **Chapter 14** Videos			
	4	Complete **Chapter 14** Study Questions, Scripture, and Spiritual Life Application			
	5				
Week 3	1	Complete **Chapter 14** "Do" Activities and Research, Observation, and Dissection			
	2				
	3				
	4				
	5				
Week 4	1	Read **Chapter 15** Pray and Sing Hymn			
	2				
	3	Watch **Chapter 15** Videos			
	4	Complete **Chapter 15** Study Questions, Scripture, and Spiritual Life Application			
	5				
Week 5	1	Complete **Chapter 15** "Do" Activities and Observation or Dissection			
	2				
	3				
	4				
	5				

GOD MADE ANIMALS - WORKBOOK

Date	Day	Assignment	Due Date	✔	Grade
Week 6	1	Read **Chapter 16** Pray and Sing Hymn			
	2				
	3	Watch **Chapter 16** Videos			
	4	Complete **Chapter 16** Study Questions, Scripture, and Spiritual Life Application			
	5				
Week 7	1	Complete **Chapter 16** "Do" Activities and Observation or Dissection			
	2				
	3				
	4				
	5				
Week 8	1	Flex Week			
	2				
	3				
	4				
	5				
Week 9	1	Complete **Final Examination**			
	2				
	3				
	4				
	5				
		Final Grade			

LESSON SCHEDULE

Chapter 1
SCIENCE FROM A BIBLICAL WORLDVIEW

STUDY QUESTIONS

Enter the name of the Christian-influenced scientist from the list of names below beside the scientific field he is known to have developed.

Louis Pasteur (1822-1895)
Isaac Newton (1642-1727, appears twice)
Robert Boyle (1627-1691)
Charles Babbage (1792-1871)
James Clerk Maxwell (1831-1879)
Michael Faraday (1791-1867)
Ambrose Fleming (1849-1945
George Stokes (1819-1903)
Johannes Kepler (1571-1630)
Lord Kelvin (1824-1907)
Carolus Linnaeus (1707-1778)

☺ great job!

1. Electronics
   John Fleming (1849-1945)
2. Thermodynamics
   Lord Kelvin (1642-1727)
3. Dynamics
   Isaac Newton ~~(1849-1945)~~ (1642-1727)
4. Chemistry
   Robert Boyle (1627-1691)

23

5. Fluid mechanics
   George Stokes (1819-1903) *Good great job.*
6. Computer science
   Charles Babbage (1792-1871)
7. Bacteriology
   Louis Pasteur (1822-1895)
8. Electromagnetics
   Michael Faraday (1791-1867)
9. What or who is the cause of everything in the Universe?

10. What do unbelieving scientists think brought about the complexity and variety of creatures in the earth?

11. What do you learn about God when you see something really big like a mountain, a lion, or a galaxy in the heavens?

12. What do you learn about God when you see something really complicated like a living cell or a spider spinning a web?

CHAPTER 1 - STUDY QUESTIONS

13. What do you learn about God when you see thousands of reindeer feeding on grasses in northern Canada, or when you see a mama bird feeding her young in a nest?

14. Man learned how to fly an airplane in the year 1903, when Wilbur and Orville Wright sustained a flight in Kitty Hawk. Who knew all about flight even before man learned about it?

15. If a grandfather hid treasures in the backyard, why would his grandchildren run out to find them? Give three or four reasons.

16. What two things did God promise Noah in Genesis 8:20-22?

GOD MADE ANIMALS - WORKBOOK

17. Why don't unbelieving scientists sing hymns in their science classes?

_____

_____

_____

_____

_____

_____

18. Give three or four purposes for our study of science that are different from unbelieving scientists.

_____

_____

_____

_____

_____

_____

SCRIPTURE

Fill in the key parts of the following scripture passages:

1. "The heavens declare the _____; and the firmament shows _____. Day unto day utters speech, and night unto night reveals knowledge. There is no speech nor language _____." (Psalm 19)

2. "So God created man in His own image; in the image of God He created him; male and female He created them. Then God blessed them, and God said to them, 'Be _____; fill the earth and subdue it; have dominion over _____ over _____, and over every living thing that_____.'"
(Genesis 1)

CHAPTER 1 - SPIRITUAL LIFE APPLICATION

3. "Then the LORD said in His heart, 'I will never again curse the ground for man's sake, although the imagination of man's heart is evil from his youth; nor will I again destroy every living thing as I have done. While the earth remains, _____, cold and heat, winter and summer, and _____.'" (Genesis 8)

4. "It is the glory of God to _____, but the glory of kings to _____." (Proverbs 25)

SPIRITUAL LIFE APPLICATION

1. List three amazing animals, natural processes, or events that you have witnessed which would cause you to praise God.

   I praise God for _____.
   I praise God for _____.
   And, I praise God for _____.

2. List some ways in which science has helped your mother in the kitchen, in very practical ways. What is it specifically that she is ruling over in her work?

   a. First way science helps mom: _____
   What is she ruling over? _____
   b. Second way science helps mom: _____
   What is she ruling over? _____

OBSERVATION OR DISSECTION

Observation is the basic method of science. This is what scientists do. But, even more importantly, this is what Christians do. We consider. We study the works of God, so that we can give Him the praise and the glory for what He has done. This applies to His work of creation, His works of providence in history, and His work of redemption by Jesus Christ who died on the cross for our sins and rose again the third day.

GOD MADE ANIMALS - WORKBOOK

"I will meditate on the glorious splendor of Your majesty, and on Your wondrous works. Men shall speak of the might of Your awesome acts, and I will declare Your greatness." (Ps. 145:5-6)

When we study what God has done in creation, we can feel, taste, listen, look, and smell. Use your senses to investigate God's world. Choose one of the following exercises as you investigate God's world for yourself.

1. Check out a dog or cat's vital signs.

   a. Use a stethoscope or you fingers, and attempt to feel or hear the heartbeat of a dog or cat.

   Hold your hand over the cat or dog's left side, just behind its front leg.

   How many beats in 10 seconds? Multiply that by six to get the number of beats per minute.

	Beats in 10 seconds	Beats in a Minute
Animal		
You		

   b. Check out a dog or cat's respiratory rate. How many breaths does he take in a minute? Just watch its chest go up and down, in and out. Compare it to your own.

   Animal respiratory rate: _____
   Your respiratory rate: _____

2. Observe the behavior of ants.

   a. How fast do they run? Time an ant as it runs for about a foot, and determine how many feet per seconds it runs.
   _____ feet/second

   b. How much distance does the ant make in 10 minutes?
   _____
   _____
   _____

CHAPTER 1 - OBSERVATION OR DISSECTION

c. How much weight can an ant carry, from your observation?
_____
_____
_____

d. Put 2-4 different kinds of food in front of it: meat, vegetable, bread, sugar. Which does the ant prefer?
_____
_____
_____
_____

3. Observe a bug or insect of any kind.

   Count its legs _____

   Count its antennae _____

   Count its body segments _____

   Where are its eyes located? _____

4. Listen to bird calls/songs. Write down the various types of calls you hear. What are the conditions under which the bird makes these calls? Is it danger? Is it communicating with another bird? Is it identifying food sources?
_____
_____
_____
_____

Chapter 2
GOD'S AMAZING CREATIVE POWER

STUDY QUESTIONS

What did God create on each of the days of creation?

1. Day 1 _____
2. Day 2 _____
3. Day 3 _____
4. Day 4 _____
5. Day 5 _____
6. Day 6 _____

7. If worldly scientists do not believe that a personal God created animals, how did it all come about, from their perspective?
   _____
   _____
   _____

8. From what raw materials do humans make the following products?

   Tables _____
   Roads and highways _____
   Tires _____
   Car hoods _____

9. From what did God make the world and all of the animals?
   _____

10. What did Jesus make the wine out of the at the wedding of Cana?

11. What are some things that Jesus accomplishes with His voice?

12. What is the one thing God did not create out of nothing (ex nihilo), when He made the world? What did He use?

13. Give several examples of things God created with the appearance of age.

CHAPTER 2 - SCRIPTURE

14. Which came first, the chicken or the egg?

_____
_____
_____
_____

15. How many fish and loaves did Jesus multiply to feed the 5,000 in John 6:1-14? Why was this an act of creation?

_____
_____
_____
_____

SCRIPTURE

Fill in the key parts of the following scripture passages:

1. "Let them praise the name of the LORD. for _____, and they were created." (Psalm 148)

2. "By _____ the heavens were made, and all the host of them by _____." (Psalm 33)

3. "Then God said, 'Let the waters abound with _____ _____, and let birds fly above the earth across the face of the firmament of the heavens.' So God created great sea creatures and every living thing that moves, _____ _____, according to their kind, and every winged bird according to its kind. And God saw that it was good." (Genesis 1)

GOD MADE ANIMALS - WORKBOOK

SPIRITUAL LIFE APPLICATION

1. What are your questions about yourself and the world around you that are answered in the Word of God? Give several examples.

   _____
   _____
   _____
   _____
   _____

2. Look around you and identify several things in your home or outdoors which man made out of materials that God made. List them here. Take a moment and praise God for providing the materials to make these things.

   _____
   _____
   _____
   _____
   _____

3. Read Exodus 20:9-11. Explain why the Bible does not support evolution — that the animals appeared over a process of millions of years.

   _____
   _____
   _____
   _____
   _____
   _____
   _____
   _____

CHAPTER 2 - OBSERVATION OR DISSECTION

4. Read Hebrews 11:3. According to this verse, what is required of you as you think about the way the earth came about at the beginning?

_____
_____
_____
_____
_____
_____

OBSERVATION OR DISSECTION

Choose one of the following exercises as you investigate God's world for yourself.

1. Go outside and observe 25 plants, animals, and inanimate objects around you that God has made. Do your best to identify each object. Then, list 10 functions or processes that God has made (wind, an animal eating, an animal flying, etc.) Acknowledge God as creator of all these things and praise Him for it.

Objects	Process/Function

2. Study the age of the living things around you. Observe a log already cut. Count the rings. Observe a flower or a plant. Observe a vertebrate animal such as a dog, a cat, or a bird. Observe an insect. What is its size? What is its age? Is it full grown or will it continue to grow? You may need to research the lifespan of certain flowers or insects to get close to the right age.

CHAPTER 2 - OBSERVATION OR DISSECTION

	Age	**Size**	**Full Grown?**
You			
Tree Log			
Flower or Plant			
Animal			
Insect			

Chapter 3
GOD MADE ANIMALS

STUDY QUESTIONS

1. What are the differences between a mechanical robot dog and a real dog?
   _____
   _____
   _____
   _____
   _____

2. "Different 'kinds' of animals created by God cannot do what?
   _____
   _____
   _____

3. Give examples for each of the following five categories of God's living creation:

   a. Category Monera _____
   b. Category Protista _____
   c. Category Fungi _____
   d. Category Plantae _____
   e. Category Animalia _____
   f. Category Man _____

GOD MADE ANIMALS - WORKBOOK

4. How is man different from the animals?

   _____

   _____

   _____

   _____

5. What are some of the similarities between man and the animals?

   _____

   _____

   _____

   _____

6. Fill in the word that fits the description from the list of words contained below:

   Genetic Code (DNA)
   Cytoplasm
   Cell
   Reproductive Systems
   Species

   a. _____A certain variety of animal that is described by its features, but may be capable of reproducing with another species.

   b. _____The gooey stuff inside of the cell which makes it alive.

   c. _____The smallest living thing named after the place where Christian monks gather to pray.

   d. _____A collection of instructions that comes from the dad dog and the mom dog which determines what the baby dog is going to look like.

   e. _____The function of a living organism to produce offspring or babies.

CHAPTER 3 - SCRIPTURE

7. Which of the following animals can interbreed, and what is the resulting animal?

   a. Donkey and Zebra _____
   b. Coyote and Dog _____
   c. Lion and Giraffe _____
   d. Cocker Spaniel and Poodle _____
   e. Wolf and Dog _____
   f. Horse and Donkey _____
   g. Raccoon and Porcupine _____

8. Which animals ate meat before the fall of Adam into sin?

   _____
   _____
   _____
   _____

9. Which animals began eating meat after the fall?

   _____
   _____
   _____
   _____

10. When did humans begin to eat meat?

    _____
    _____
    _____
    _____

GOD MADE ANIMALS - WORKBOOK

SCRIPTURE

Fill in the key parts of the following scripture passages:

1. "Then God said, "Let Us make man in _____, according to _____; let them have dominion over the fish of the sea, over the birds of the air, and over the cattle, over all the earth and over every creeping thing that creeps on the earth." So God created man _____ He created him; male and female He created them. (Genesis 1)

2. "What is _____ that You are mindful of him, and the _____ that You visit him? For You have made him a little lower than the angels, and You have crowned him with glory and honor. You have _____; you have put _____ _____ . . ." (Psalm 8)

3. "Out of the ground the LORD God formed every beast of the field and every bird of the air, and brought them to Adam to see what _____. And whatever Adam called each living creature, that was its _____. So Adam gave _____ _____, to the birds of the air, and to every beast of the field. But for Adam there was not found a _____ to him." (Genesis 2)

4. "For we know that the whole creation _____ together until now. Not only that, but we also who have the firstfruits of the Spirit, even we ourselves groan within ourselves, eagerly waiting for the adoption, _____. (Romans 8)

CHAPTER 3 - SPIRITUAL LIFE APPLICATION

SPIRITUAL LIFE APPLICATION

1. Study Psalm 8. What is the hierarchy (order) of dignity and honor in God's creation?

   God

   _____

   _____

   _____

   What are the parts of creation we (humans) are called to rule over?

   _____

2. Compare yourself to your mother and father — your personality, your looks, your size and shape. Which characteristics did you get from your father? Which did you get from your mother? (If you only have known one parent, just include that information below.) Take a moment and thank God for these characteristics that He gave you from your parents.

   _____
   _____
   _____
   _____
   _____
   _____
   _____
   _____
   _____
   _____

GOD MADE ANIMALS - WORKBOOK

OBSERVATION OR DISSECTION

Choose one of the following exercises as you investigate God's world for yourself.

1. Observe a cockapoo, a puggle, and a labradoodle. Which characteristics do these dogs receive from each parent?

   Cocker Spaniel characteristics: _____
   Poodle characteristic: _____

   Pug characteristics: _____
   Beagle characteristics: _____

   Poodle characteristics: _____
   Labrador characteristics: _____

2. Observe and record the eye color of ten children, from at least five different families. Fill in the color of their parent's eyes as well.

   The brown eye gene is dominant over the blue eye gene. That means the brown eye gene is more influential than the blue eye gene. The blue eye gene is recessive, which means it has a weaker influence on whether the children will have the same gene.

   So, if a child has a parent with brown eyes and a parent with blue eyes, the child will have brown eyes most of the time. If both parents have blue eyes, the children will have blue eyes most of the time.

	Name	Eye Color	Parent 1 Eye Color	Parent 2 Eye Color
Child #1				
Child #2				
Child #3				
Child #4				
Child #5				

Child #6				
Child #7				
Child #8				
Child #9				
Child #10				

What kind of pattern do you see in this?

_____

_____

_____

_____

_____

_____

Chapter 4
ANIMALS WITHOUT BACKBONES

STUDY QUESTIONS

1. Fill in the word that best fits each phrase using the list of words contained below:

Coral
Metamorphosis
Insect
Bivalve mollusks
Arachnid
Arthropod
Cephalopod
Crustacean
Invertebrate
Gastropods

a. _____ An invertebrate with the head and foot combined in one organ

b. _____ An animal that is sensitive to ocean water temperatures

c. _____ Animals without backbones

d. _____ Invertebrates with one foot (containing a stomach)

e. _____ Invertebrates with a segmented body and jointed appendages

f. _____ Clams, oysters, scallops

GOD MADE ANIMALS - WORKBOOK

g. _____ The process of changing form during a lifespan

h. _____ An 8-legged arthropod

i. _____ A 10-legged arthropod

j. _____ A 6-legged arthropod

2. What is the most dangerous spider in the world?
_____
_____
_____

3. Why are ticks so dangerous?
_____
_____
_____
_____

4. What is a chigger? Why is it so harmful?
_____
_____
_____
_____

5. Give an example of each of the following animals:

a. Insect
_____

b. Bivalve mollusk
_____

c. Univalve mollusk
_____

CHAPTER 4 - STUDY QUESTIONS

d. Gastropod
_____

e. Cephalopod
_____

f. Arachnid
_____

g. Animals that undergo metamorphosis
_____

h. Crustacean
_____

6. What wonderful defense mechanism has the Lord provided the following creatures?

Octopus _____
Mollusks _____
Jellyfish _____
Flea _____
Bombardier Beetle _____
Spiders _____

7. What is the strongest animal in the world, ounce for ounce?
_____
_____
_____

8. What is the difference between a butterfly and a moth?
_____
_____
_____
_____
_____

GOD MADE ANIMALS - WORKBOOK

9. Describe the four stages of the butterfly using the following clues:

   a. Egg
   b. C _____
   c. C _____
   d. B _____

10. What are the character traits of the ant?
    _____
    _____
    _____
    _____
    _____

11. How does the Lord benefit mankind and the earth using the following animals?

    Earthworms
    _____

    Crustaceans
    _____

    Spiders
    _____

12. What are the animals that are generally seen as pests, covered in this chapter?
    _____
    _____
    _____
    _____
    _____

13. Name that spider. To which spider has God given each of the amazing qualities listed below:

    a. This spider hides under a trap door of its own making and waits to ambush a victim.

50

CHAPTER 4 - SCRIPTURE

_____

b. This spider is huge—it could fill your whole dinner plate. _____

c. This spider is able to live under water by building a diving bell out of its silk.
_____

d. This spider builds a net out of its web and uses it to capture its prey.
_____

e. This spider can set up a web across a river some 80 feet (25 m) wide.
_____

SCRIPTURE

1. Praise the LORD from the earth, you _____ and all the depths; fire and hail, snow and clouds; stormy wind, fulfilling His word; mountains and all hills; fruitful trees and all cedars; beasts and all cattle; _____ and flying fowl. (Psalm 148)

2. The LORD is gracious and full of compassion, slow to anger and great in mercy. The LORD is good to _____, and His tender mercies are over _____. (Psalm 145)

3. This great and wide _____, In which are _____, Living things both small and great. There the ships sail about; There is that _____ Which You have made to play there. These all wait for You, That You may give them _____ _____. (Psalm 104)

4. There are...things which are _____, But they are exceedingly wise: The _____ not strong, Yet they prepare their food in the summer. (Proverbs 30)

GOD MADE ANIMALS - WORKBOOK

5. There be... things which are little upon the earth, but they are exceeding wise: The _____ _____, and is in kings' palaces. (Proverbs 30:24, 28 KJV)

SPIRITUAL LIFE APPLICATION

1. Study Psalm 148:7-11. What might the Lord be referring to, when He speaks of "great sea creatures" praising Him? What about creeping things? What could He be referring to here?

   _____
   _____
   _____
   _____
   _____
   _____
   _____
   _____
   _____

2. What is the most amazing animal you studied in this chapter? What is your favorite, and what attributes of God does this display?

   _____
   _____
   _____
   _____
   _____
   _____
   _____
   _____
   _____

CHAPTER 4 - OBSERVATION AND DISSECTION

OBSERVATION AND DISSECTION

Choose one of the following activities below and complete the exercises:

1. Watch the miracle of metamorphosis. Find a caterpillar or order one or two. The caterpillar (larvae) stage and the pupa stage will last about two weeks each.

   Follow directions here, using this helpful website: https://learning-center.homesciencetools.com/article/butterfly-science-projects-and-activities/

   Order caterpillars here: https://www.homesciencetools.com/product/insect-lore-butterfly-garden-kit/

   Observe especially the following wonders God worked into this beautiful creation of His.

   a. Note the symmetry in the butterfly design. Our God's design of beauty involves the same appearance on one side of the animal as on the other side. Describe the pattern of the wings and other parts of the insect on one side and on the other side.

   b. Count the legs on the caterpillar. It should have 3 pair of "pro-legs" up front, which qualifies the little guy as an insect.

   c. Note the difference in what the caterpillar and the butterfly will eat. Also note how the insect eats.

   d. Praise the Lord for the miracle of metamorphosis! Consider the power of God in creation. What you see cannot be done by human science and human technology. Only God can produce such a wondrous thing.

2. Dig up an earthworm.

   a. Run your fingers up and down the body of the worm. Do you feel the tiny sharp prickly hairs? These are called "setae" and they keep the earthworm secure in the dirt. If a bird tries to pull it out of the dirt, the setae keeps it stuck down in there. Sometimes, the bird will break off a part of the worm and leave the rest in the dirt. The setae also help the worm gain some traction while scooching along the ground.

GOD MADE ANIMALS - WORKBOOK

Take a moment, and praise the Lord for providing this design as a way of protecting this little invertebrate.

b. Count the hearts (near the mouth of the earthworm) on the end where you will find a swollen band. You should count 5 pairs of hearts.

c. Cut a part of the worm's tail off. Examine the blood. What color is it? Keep the earthworm in a jar for a week or two, and watch to see if the tail will grow back.

Note: A planarium flatworm can grow a new head on a segment of its body, and believe it or not, its new head can retain old memories as well.

3. Dissect an earthworm using the instructions available on this website: https://learning-center.homesciencetools.com/article/earthworm-dissection/

Acknowledge God's wisdom in creating a digestive system for the earthworm that can process dirt and still retain some nutrition! The earthworm is God's cleanup crew under the ground, and this creature helps the dirt to produce crops, by breaking up the ground. Praise God for His wisdom and goodness in the creation of the earthworm!

4. Obtain a snail from the garden or somewhere else. Use a magnifying glass to study the snail. Handle your snail with plastic gloves (or be sure to wash your hands after touching it.)

    a. How many tentacles can you see? _____
    b. Where are the snail's eyes located? _____
    c. Where is the snail's foot?_____
    d. Place your snail on a piece of transparent smooth plastic. Watch from underneath using your magnifying glass to see how it can move along without sliding backwards every time its foot moves forward.

    _____

    _____

    _____

    e. Allow your snail to crawl around for a while in a box lined with a dark piece of paper. When it is done, sprinkle the paper with white talcum powder to follow its tracks. How does it move? Does he move in a straight line?

    _____

f. Time your snail's movements. How many seconds does it take it to move about one foot? Take 0.68 and divide that by the number of seconds, and that is how many miles per hour your snail can move.

g. Place a lid over 3/4 of the box opening. Does the snail move towards the light or away from the light?

Chapter 5
HOW INSECTS SERVE GOD AND HIS CREATION

STUDY QUESTIONS

1. What are God's animals He uses to clean up things around the world?

    a. Cleaning up on the ocean floor _____
    b. Cleaning up under the soil _____
    c. Cleaning up dung on top of the soil _____

2. What are animals that eat insects called?
_____
_____
_____

3. What are animals that eat meat called?
_____
_____
_____

4. What are animals that eat plants called?
_____
_____
_____
_____

GOD MADE ANIMALS - WORKBOOK

5. What are animals that eat just about anything called?

6. Why is the mosquito the most dangerous animal in the world?

7. How are the fly's eyes better than ours?

8. How do flies pass disease around?

9. What diseases to ticks and lice communicate?

10. What kind of insects are the worst pest for farmers?

CHAPTER 5 - STUDY QUESTIONS

11. Which insects are a farmer's friends?

_____

_____

_____

_____

12. Where is the pollen (or the male cell) in a plant?

_____

_____

_____

13. Where is the female cell located?

_____

_____

_____

14. How do bees help the pollen find the female cells?

_____

_____

_____

15. Name the four things the leech's saliva (as God invented it) can provide, which man has used for medical applications.

_____ To help prevent blood clotting

_____ To help numb some area of the body

_____ To open up blood vessels to help the blood flow better

_____ A medicine to help attack certain bacterial infections

59

GOD MADE ANIMALS - WORKBOOK

SCRIPTURE

1. Then God spoke to Noah, saying, "Go out of the ark, you and your wife, and your sons and your sons' wives with you. Bring out with you every living thing of all flesh that is with you: birds and cattle and every _____, so that they may _____ on the earth, and be _____ on the earth." (Genesis 8)

2. The fear of the LORD is clean, enduring forever; _____ are true and righteous altogether. More to be desired are they than gold, yea, than much fine gold; sweeter also than _____. Moreover by them Your servant is warned, and in keeping them there is great reward. (Psalm 19)

3. He causes the grass to grow for the cattle, And vegetation for _____ _____, That he may bring forth _____, And _____ _____, _____ to make his face shine, And _____ which strengthens man's heart. The trees of the LORD are full of sap, The cedars of Lebanon which He planted, Where the birds make their nests; The stork has her home in the fir trees. (Psalm 104)

SPIRITUAL LIFE APPLICATION

1. Find a few more examples of how much God appreciates reproduction and provides many opportunities for it in His creation. Look at the plants and look at certain animals. How much pollen can you see in a flower? How many seeds come out of certain plants? How many eggs are laid by certain fish or frogs? Be sure to praise Him and thank Him for the many ways in which He keeps life going on all around us. Note several examples in the space provided below:

_____

_____

_____
_____
_____
_____
_____
_____

_____

2. From Psalm 104, list all of the amazing things the Lord has provided man out of His creation.

_____
_____
_____
_____
_____
_____
_____
_____
_____
_____
_____
_____
_____
_____

GOD MADE ANIMALS - WORKBOOK

OBSERVATION AND DISSECTION

Choose one of the following activities below and complete the exercises.

1. Study the scout bee's little wiggle dance. Draw out two different patterns, and indicate the direction and location of the pollen source.

   Pattern 1

   Pattern 2

2. Test the strength of a spider web by hanging little weights on the spider web until it breaks. Watch the assigned video on the *God Made Animals* webpage that compares the strength of the spider web to the strength of steel.

   What kind of spider made the web that you studied?
   _____
   _____

   How much weight did it take to break the spider web? How many little weights did you have to put on the web to break it?
   _____
   _____

CHAPTER 5 - OBSERVATION AND DISSECTION

3. Visit some bee hives, and add to the following fun facts about bees:

- The queen bee can lay up to 2,500 eggs in one day.
- A bee can travel a distance as far as 1 1/2 times around the circumference of the earth in her lifespan of 6-8 weeks.
- On a single trip, the worker bee can hit up to 100 flowers. That's why they are busy as bees!
- Honey is about 25% sweeter than regular white sugar.
- Bee pollen is a super food, and our Creator has put in it practically all the nutrients needed to sustain life.
- Honeybee venom is more poisonous than the Cobra's, but it doesn't hold enough to do much damage to a human.
- Honey made from the Rhododendron flower is poisonous.

Be prepared to ask good questions.

What sort of plants produce the best honey?

_____
_____
_____
_____
_____

What are the things that attract bees?

_____
_____
_____
_____

How do you avoid getting stung by a bee?

_____
_____

Chapter 6

GOD MADE BIRDS

STUDY QUESTIONS

1. What are the three different environments God has made in which life exists?
   _____
   _____
   _____
   _____

2. What sorts of animals exist in extreme cold?
   _____
   _____
   _____
   _____

3. What sort of life exists above 20,000 feet?
   _____
   _____
   _____

4. What passages in both the Old Testament and the New Testament forbid the eating of blood?
   _____
   _____

GOD MADE ANIMALS - WORKBOOK

5. Provide examples of animals with blood and animals without blood.

6. What is the difference between warm-blooded animals and cold-blooded animals?

7. Provide examples of warm-blooded animals and cold-blooded animals.

8. How do dogs and birds cool off their bodies on hot days? How did God design your body to cool itself off?

9. Which animals will hibernate in very cold weather?

CHAPTER 6 - SCRIPTURE

10. How do birds break down their food for digestion? Provide the two-step process.

11. World record holders. Name the bird that...

    a. Flies the farthest for its migration

    b. Dives as deep as 250 feet to grab a fish

    c. Hibernates in the western United States

    d. Flies as high as 23,000 feet

SCRIPTURE

1. Every moving thing that lives shall be food for you. I have given you all things, even as the green herbs. But you shall not eat _____, that is, its _____. (Genesis 9)

2. For He says to the _____, "Fall on the earth;" Likewise to the gentle rain and the heavy rain of His strength. He seals the hand of every man, That all men may know His work. The _____, And remain in their _____. (Job 37)

GOD MADE ANIMALS - WORKBOOK

3. But those who wait on the _____ Shall renew their strength; They shall mount up with _____, They shall run and not be _____, They shall walk and not _____. (Isaiah 40)

4. "Therefore I say to you, do not worry about your life, _____ _____; nor about your body, what you will put on. Is not life more than food and the body more than clothing? Look at _____, for they neither sow nor reap nor gather into barns; yet your heavenly Father feeds them. Are you not of more value than they?" (Matthew 6)

SPIRITUAL LIFE APPLICATION

1. Study Matthew 6:25-32. If humans take care of and feed about 10 billion chickens and 5 billion cows, sheep, goats, dogs, and cats, how does that compare to all of the mammals, fish, insects, and birds that God feeds? Have you ever been tempted to worry? Why should you not worry about your material needs?

_____
_____
_____
_____
_____

2. Study Isaiah 40:31. Have you ever been very tired and you did not feel like you could go on? What lesson does the Arctic tern and the bar-tailed godwit provide for us? Where do they get their strength?

_____
_____
_____
_____
_____

CHAPTER 6 - OBSERVATION OR DISSECTION

OBSERVATION OR DISSECTION

Choose one (or preferably two) of the following exercises as you investigate God's world for yourself.

1. Watch the birds in your backyard for about 30 minutes. Birds tend to come out directly after sunrise and before sunset. If you have a camera, take pictures of each of the birds. Try to identify the birds. What sorts of sounds do they make? What do they eat? Are they helpful or hurtful to your yard and your home?

   _____
   _____
   _____
   _____
   _____
   _____

   Record the shapes, sizes, colors, and patterns of each bird.

   _____
   _____
   _____
   _____

   Praise God for each of the birds you identified.

2. Dissect an insect, preferably a grasshopper. Can you identify the sac containing the hemolymph? What color is the hemolymph? What is its consistency? Is it sticky or does it flow freely?

   _____
   _____
   _____
   _____

GOD MADE ANIMALS - WORKBOOK

Use the following website to guide you through the dissection and observation: https://www.biologyjunction.com/grasshopper_dissection.htm

3. Obtain some samples of goose, duck, and chicken feathers from a farm (or online). Compare the softness of these feathers and the insulate qualities (to protect you from the cold). Compare these with man-made pillows and insulation materials. Place the same amount of insulation around your finger and touch an ice cube. Which material provides the best insulation for your finger from the cold?

Take a moment to praise the Lord for the comfort and the protection from the cold provided by the duck, the goose, and the chicken.

4. Touch both your hands to your cheeks to make sure they are the same temperature. Submerge one hand in room temperature water. Twirl your arms about like windmills until your wet hand is dry. Place both hands on your cheeks again. What do you observe? How does sweat cool off the body?

Prepare a bowl of ice water. Coat the index finger of your right hand with cooled meat grease, shortening, or ice water. Leave your left index finger clean. Place both fingers into the ice water. What did you observe? How do you think layers of fat on hairless animals might help them in cold water or cold weather?

5. On a clear day, go out an hour or two after sunrise and put your hand on the east wall of a building. Then put your hand on the west wall. What is the difference in temperature? Do the same in the late afternoon. What is the difference in temperature at this time of day? How might a cold blooded animal take advantage of the principle you have discovered?

Chapter 7
GOD'S AWE-INSPIRING DESIGN FOR BIRD FUNCTION

STUDY QUESTIONS

1. What did Orville and Wilbur Wright do first as they set out to develop the first airplane?

2. What are some of the design features God put into the bird to enable flight?

3. From the list of birds contained below, identify the bird that best fits the description provided.

Woodpecker
New Caledonian crow
Parrot

GOD MADE ANIMALS - WORKBOOK

Peregrine falcon
African hamerkops
Golden eagle
Chimney swift

_____ Uses its own spittle for glue to hold its nest together

_____ Builds the largest nest in the world

_____ Can learn to speak as many as 100 words

_____ Fashions a tool to get grubs out of a tree

_____ Uses 12,000 twigs and water proofs and insulates its nest on the inside and out with mud.

_____ Beak is built to withstand incredible shock

_____ Fastest animal in the world, flying at 240 mph

4. Fill in the emperor penguin's reproductive cycle by the following months:

   January-March _____
   April-May _____
   June-July _____
   August _____
   September-October _____
   October-December _____

5. What kind of celebratory antics do the following birds produce when they are getting ready to reproduce?

   a. Satin bower bird

   _____

   _____

   b. Hummingbird

   _____

c. Great crested grebe

d. Puffin

6. Provide several examples of male/female birds that will stay together for life.

SCRIPTURE

1. He sends _____; they flow among the hills. They give drink to every beast of the field; The wild donkeys quench their thirst. By them _____ _____; they _____ among the branches. He waters the hills from His upper chambers; The earth is satisfied with the fruit of Your works. (Psalm 104)

2. But now ask the beasts, and they will teach you; and the_____ and they will tell you; or speak to the earth, and it will teach you; and the fish of the sea will explain to you. Who among all these does not know that _____ _____ has done this, in whose hand is the_____, and the breath of all mankind? (Job 12)

GOD MADE ANIMALS - WORKBOOK

3. Even the _____ has found a home, And the swallow _____ _____, Where she may lay her young— Even Your _____, O LORD of hosts, My King and my God. Blessed are those who _____; They will still be praising You." (Psalm 84)

SPIRITUAL LIFE APPLICATION

1. 1. Study Psalm 104:10-13. How do the springs of the valley provide the birds of the heavens a home? How has God provided you a home? What is the home built out of? Which of God's materials did somebody use to build your home?

   _____
   _____
   _____
   _____
   _____
   _____

2. Where have the birds found a home in Psalm 84? What kind of lesson does this teach you relating to God's house—the church? Do you share the psalmist's feelings towards the church (contained in verses 1-4)?

   _____
   _____
   _____
   _____
   _____
   _____
   _____

CHAPTER 7 - OBSERVATION OR DISSECTION

OBSERVATION OR DISSECTION

Choose at least one (preferably two) of the following exercises as you investigate God's world for yourself.

1. Search in the bushes and trees around your house (or classroom) for an abandoned bird's nest.

   a. Record the size and weight of the nest.
   Size: _____
   Weight: _____

   b. How was the nest secured to the tree, bush, or place where you found it?
   _____
   _____

   c. Estimate about how many twigs, grass blades, or other objects it took to build the nest. Where did the bird find all of these materials?
   _____
   _____
   _____

   d. Try to determine how the nest was knitted or kept together. What was the technique used?
   _____
   _____
   _____

   e. By examining the nest and the birds in the area, can you guess what kind of bird built the nest?
   _____
   _____
   _____

GOD MADE ANIMALS - WORKBOOK

f. Praise God for the ingenuity and skill He put into the birds to build such a nest as this.

2. Study the design God brought about in the birds. Using some of the features, design your own airplane out of thin paper or cellophane for maximum lift. Find a way to make an extremely light airplane while providing it some structure. Experiment with two or three designs, and explain why one design is better than the other.

3. Observe birds as they eat at your feeder or somewhere else. How many times a minute can you see them pecking food off the ground? Assuming that a bird is doing just that for twelve hours at a time, estimate how many seeds they would consume in that time period.

Chapter 8
GOD'S COMMAND TO RULE OVER THE BIRDS

STUDY QUESTIONS

1. Fill in the blanks provided below with the term "helpful" or "harmful," to best describe this animal's contribution to life on earth for man.

    a. Honey bee _____
    b. Vulture _____
    c. Butterfly _____
    d. Slug _____
    e. Dung beetle _____
    f. Ladybug _____
    g. Aphid _____
    h. Tick _____
    i. Japanese beetle _____
    j. Chicken_____
    k. Mosquito_____

2. What type of animal spreads the following diseases?

    a. Rabies _____
    b. Tularemia bacteria _____
    c. Bubonic plague_____

3. List at least three domesticated birds and mammals.

    a. _____

GOD MADE ANIMALS - WORKBOOK

b. _____
c. _____

4. What are the two uses of pigeons?
_____
_____
_____
_____

5. What are three benefits of chickens?
_____
_____
_____
_____
_____
_____

SCRIPTURE

1. Our soul has escaped as a _____ of the fowlers; the snare is broken, and we have escaped. Our help is in _____, who made heaven and earth. (Psalm 124)

2. Be diligent to know the _____. And attend to your _____; for riches are not forever, nor does a crown endure _____ _____. (Proverbs 27)

3. Now Abel was a _____, but Cain was a tiller of the ground. And in

82

the process of time it came to pass that Cain brought an offering of the fruit of the ground to the LORD. Abel also brought of the _____ and of their fat. And_____. (Genesis 4)

SPIRITUAL LIFE APPLICATION

1. How have you escaped the snare of the fowler? Has the Lord helped you escape from certain sins and temptations in your life? Note them below.

_____
_____
_____
_____
_____
_____
_____
_____

2. How much chicken does your family eat in a given year? How many eggs does your family consume in a year? How does this compare to the amount of beef or other meat you eat? Given that number, try to calculate how many chickens are consumed in your town in a given year. Take a moment to thank the Lord for all of that food.

_____
_____
_____
_____
_____
_____
_____
_____
_____

GOD MADE ANIMALS - WORKBOOK

OBSERVATION OR DISSECTION

Choose at least one (preferably two) of the following exercises as you investigate God's world for yourself:

1. Research various breeds of chicken:

   a. Which breeds are the best at laying eggs (per year)?
   _____
   _____
   _____

   b. Which breeds do the best in cooler climates? And warmer climates?
   _____
   _____
   _____

   c. Which breeds grow fastest?
   _____
   _____
   _____

   d. Which breeds make the best broilers? Which provide the most meat?
   _____
   _____
   _____
   _____

2. Obtain a whole chicken from the store, and study the meat. What part of the meat is white? What part of the meat is dark?
   _____
   _____

CHAPTER 8 - OBSERVATION OR DISSECTION

_____
_____
_____

Note: Dark meat contains more myoglobin, which settles in muscles that are used more regularly. Which muscles are used more regularly in a chicken?

_____
_____
_____
_____

Weigh a chicken bone. Weigh a beef bone about the same size. What is the difference in weight?

_____
_____
_____

Cut the chicken bone in two and examine the cross section. Make a drawing of it in the space provided below. Why is the chicken bone so light?

_____
_____
_____
_____

85

Chapter 9
GOD MADE FISH

STUDY QUESTIONS

1. Fill in the blanks with the appropriate term:

    a. _____ Animals with backbones

    b. _____ Animals without backbones

    c. _____ Bones that protect the spinal cord

    d. _____ Bone that protects the brain

    e. _____ Made up of two hydrogen atoms and one oxygen atom

    f. _____ Noses and ears are made out of this

2. What body parts do fish possess and use to breathe, but that humans do not have?
   _____
   _____
   _____

3. What are the five kinds of fins on a fish?
   _____
   _____

GOD MADE ANIMALS - WORKBOOK

4. Which fin is the primary means of propelling the fish through the water?
_____
_____

5. From the list of fish contained below, identify the fish that best fits the description provided.

Black swallower
Walking catfish
Hawaii goby fish
Spotted ray
Archer fish
Marlin
Stonefish
Mandarin Fish

_____ a. Fastest swimmer in the world
_____ b. Looks just like sand and pebbles
_____ c. Most poisonous fish in the world
_____ d. Can eat a meal ten times its own size
_____ e. Uses his mouth like a squirt gun to shoot insects
_____ f. Most beautiful fish in the world
_____ g. Survives on dry land for a few days
_____ h. Rock climber

6. Match the animals that help each other in some kind of symbiosis.

    Sea Anemone   Blind pistol shrimp
    Blue streak wrasses   Sea slug
    Imperial Shrimp   Oriental sweetlips
    Goby Fish   Boxer Crab

7. What sort of defense mechanisms has the Lord designed for the following fish?

   a. Porcupine fish _____
   b. Surgeon fish _____
   c. Stingray _____

88

CHAPTER 9 - SCRIPTURE

    d. Lion fish _____
    e. Coral reef fish _____
    f. Mandarin fish _____

SCRIPTURE

1. Oh, give thanks to the LORD, for _____! For His mercy endures forever. . . Who _____ all flesh, for His mercy endures forever. (Psalm 136)

2. Then God said, "Let the waters abound _____
_____, and let birds fly above the earth across the face of the firmament of the heavens." So _____ and every living thing that moves, with which the waters abounded, according to their kind. . . (Genesis 1)

3. For the creation _____, not willingly, but because of Him who subjected it in hope; because the creation itself also will be delivered from the bondage of corruption into the glorious liberty of the children of God. For we know that _____ with birth pangs together until now. Not only that, but we also who have the first fruits of the Spirit, even we ourselves groan within ourselves, eagerly waiting for the adoption, the _____. (Romans 8)

SPIRITUAL LIFE APPLICATION

1. Study Romans 8:20-23. Have you ever had a pet die? Have you ever seen a dead animal? How did you feel about this death? Explain your feelings about the death of animals. What brought this death to the earth? How does Romans 8:23 comfort us?

_____
_____
_____

GOD MADE ANIMALS - WORKBOOK

2. What have you learned about the goodness of God in this chapter on fish? What have you learned about the wisdom of God?

OBSERVATION OR DISSECTION

Choose one of the following exercises as you investigate God's world for yourself.

1. Identify 5-10 of the most beautiful fish in the seas. Describe these fish in words or draw and color their patterns in the space provided below. Focus particularly on the variety of colors and designs. Take a moment to praise God for His beauty that He has placed in creation.

CHAPTER 9 - OBSERVATION OR DISSECTION

2. Dissect a perch or salmon. Use this guide available at: https://www.homesciencetools.com/product/perch-dissection-guide/ or http://www.pskf.ca/sd/.

   a. Draw a picture of a fish, and identify the five fins on your fish.

   b. Draw the lateral line into the picture of your fish.

   c. Identify the operculum on your fish. What does the operculum do?
   _____
   _____
   _____

   d. Identify the location of the vent.
   _____
   _____

e. Identify the relative location of the following internal organs: the heart, the gills, the swim bladder, the stomach, and the intestine.

___

___

___

___

f. Give God the glory for His amazing creation of the fish.

Chapter 10
GOD DESIGNED FISH FOR A PURPOSE

STUDY QUESTIONS

1. What animals does God use to clean up dead animals on the sea floor?

_____
_____
_____
_____
_____

2. What are three differences between how God made sharks and how He made other fish?

a. _____
b. _____
c. _____

3. From the list of fish or sea creatures contained below, identify the fish that best fits the description provided.

Irukandji box jellyfish
Great white shark
Whale shark
Hagfish
Greenland shark
Siamese fighting fish

GOD MADE ANIMALS - WORKBOOK

_____ a. Makes a nest out of bubbles from its own saliva
_____ b. Largest fish in the world
_____ c. Lives for as long as 272-512 years
_____ d. Most dangerous shark in the world
_____ e. Second most dangerous animal in the ocean
_____ f. Ugly beast without a spine, only a skull made of cartilage

4. How does a sockeye salmon find its way back to the river where it was born?
_____
_____
_____
_____

5. What does the female fish contribute and what does the male fish contribute to produce a baby fish?
_____
_____
_____
_____

SCRIPTURE

1. The LORD is my light and my salvation; whom shall I _____? The LORD is the strength of my life; of _____? When the wicked came against me to _____, my enemies and foes, they _____. (Psalm 27)

2. And do not fear _____ but cannot kill the soul. But rather fear Him who is able to destroy both soul and body in hell. Are not two sparrows sold for a copper coin? And not one of them falls to the ground apart _____

CHAPTER 10 - SPIRITUAL LIFE APPLICATION

_____. But the very hairs of your head are all numbered. Do not _____ therefore; you are of more value than many sparrows. (Matthew 10)

SPIRITUAL LIFE APPLICATION

1. Are there any animals which frighten you? List them below. What does Matthew 10:28-31 encourage you to do?

___

___

___

___

___

___

___

___

2. Suppose one of your friends is bitten by a shark or takes a serious jellyfish sting while swimming in the ocean. What would you do?

___

___

___

___

___

___

GOD MADE ANIMALS - WORKBOOK

OBSERVATION OR DISSECTION

Choose one of the following exercises as you investigate God's world for yourself.

1. Go snorkeling yourself and observe fish. Or observe underwater life on videos available on the internet (with adult supervision).

    Keep in mind that fish are more active in the morning hours than later in the day. Also, most fish congregate around rocks and coral reefs for their protection.

    a. Note how large schools of fish travel together with amazing coordination. They will appear like they move as a single organism. How do they coordinate their movement so well?

    _____
    _____
    _____
    _____

    b. What sort of fish did you observe? Describe the variety of the fish—the colors, shapes, and sizes?

    _____
    _____
    _____
    _____
    _____
    _____
    _____

    c. How does the underwater world look different than the world we live in? Describe the coral, the plant life, and other organisms you see.

    _____
    _____

_____

_____

_____

_____

_____

_____

d. Praise God for the beauty and the teeming life in the seas.

2. Take a class in First Aid or Basic Babysitting from your local Red Cross organization (or from a local ministry or hospital). Use this website to identify a class near you (if you are in the United States) — www.redcross.org/take-a-class.

Chapter 11
TAKING DOMINION OF THE FISH

STUDY QUESTIONS

1. How many fish did the disciples catch in John 21?

_____

_____

_____

2. How many pounds (or kg) of fish did Jesus make for 5,000 men, if each man ate 1/2 pound of fish?

_____

_____

_____

_____

3. How many pounds (kg) of fish do people eat every year in the entire world? How much more chicken than fish do people eat in a year?

_____

_____

_____

4. What is the most popular fish in the world for human consumption? What are the most popular fish in America?

_____

GOD MADE ANIMALS - WORKBOOK

_____

_____

5. Where are the best kept oceans in the world?

_____

_____

_____

6. Where are the worst kept oceans in the world?

_____

_____

_____

SCRIPTURE

1. What is_____ that You are mindful of him, And the son of man that You visit him? For You have made him a little lower than the _____, And You have crowned him with glory and honor. You have made him to have _____ over the works of Your hands; You have put_____, All sheep and oxen— Even the beasts of the field, The birds of the air, And the _____ That pass through the paths of the seas. (Psalm 8)

2. Then Jesus said to them, "Children, have you any food?" They answered Him, "No." And He said to them, "_____ on the right side of the boat, and you will find some." So they cast, and now they were not able to draw it in because of the _____. Therefore that disciple whom Jesus loved said to Peter, "It is the Lord!" Now when Simon Peter heard that it was the Lord, he put on his outer garment (for he had removed it), and plunged into the sea. But the other disciples came in the little boat (for they were not far from land, but about two hundred cubits), dragging the net with fish. Then, as soon as they had come to land, they saw a fire of coals there, and_____, and bread. Jesus said to them, "Bring some of the fish

which you have just caught." (John 21)

3. "There will be a _____, because these waters go there; for they will be healed, and everything will live wherever the river goes. It shall be that _____ from En Gedi to En Eglaim; they will be places for spreading their nets. Their fish will be of the same kinds as the _____ _____." (Ezekiel 47)

SPIRITUAL LIFE APPLICATION

Read the following from Matthew 4:18-22:

And Jesus, walking by the Sea of Galilee, saw two brothers, Simon called Peter, and Andrew his brother, casting a net into the sea; for they were fishermen. Then He said to them, "Follow Me, and I will make you fishers of men." They immediately left their nets and followed Him. Going on from there, He saw two other brothers, James the son of Zebedee, and John his brother, in the boat with Zebedee their father, mending their nets. He called them, and immediately they left the boat and their father, and followed Him.

1. What were the names of the four fishermen Jesus chose to be His disciples?

2. What were the men doing when Jesus came to them?

3. What are the characteristics of a fisherman?

   a. Is he optimistic or pessimistic that he will catch fish?

b. Is he patient or impatient?

c. Is a fisherman willing to suffer the cold, the heat, and the hardship, or does he avoid the cold, the heat, and the hardship?

d. Should the fisherman trust that God will help him catch the fish?

e. How would you fish for men?

OBSERVATION OR DISSECTION

Choose one of the following exercises as you investigate God's world for yourself.

1. Do some research to identify the fishing laws in your state or region.

    How much does a license cost for children? For adults?

    What are the bag limits governing the number of fish you can catch for the 3 most popular fish?

_____

_____

_____

_____

What are the most common fish in the rivers, lakes, and seas in your area?

_____

_____

_____

_____

What are the law relating to using nets? What are you allowed to catch with fishing nets?

_____

_____

_____

_____

2. Obtain a fishing rod, and work on casting in the backyard. Be sure there aren't any other people near you so you don't catch them with the hook.

   Learn how to tie a line to a hook. Use the improved clinch knot.

Chapter 12
GOD MADE AMPHIBIANS

STUDY QUESTIONS

1. Fill in the blanks provided with one or more of the following terms:

Vertebrate
Invertebrate
Arthropod
Insect
Amphibian
Arachnid
Crustacean
Metamorphosis
Gastropod
Scavenger
Cephalopod
Mollusk
Exoskeleton
Nocturnal
Cartilage
Endoskeleton

a. _____ Salmon
b. _____ Fly
c. _____ Frog
d. _____ Shrimp
e. _____ Shark

GOD MADE ANIMALS - WORKBOOK

    f. _____ Grasshopper
    g. _____ Jellyfish
    h. _____ Vulture
    i. _____ Butterfly
    j. _____ Tick
    k. _____ Snail
    l. _____ Octopus
    m. _____ Trapdoor Spider
    n. _____ Centipede

2. How does the tadpole differ from the frog?

_____

_____

_____

_____

3. Outline the stages of the frog's life (from egg to death).

_____

_____

_____

_____

_____

4. Explain what lateral lines do.

_____

_____

_____

_____

5. What happens to the stickiness of the frog's saliva between catching the insect and swallowing?

_____

_____

CHAPTER 12 - SCRIPTURE

6. Fill in the blanks with the amphibian that best fits the description using the list provided.

Bull frog
Wood frog
Glass frog
Poison frogs
Golden frog
Darwin frog
Salamander
African clawed frog

a. _____ Parents carry their tadpoles on their backs.
b. _____ Cycle from egg to frog takes three years.
c. _____ Male frog keeps babies in vocal sac for six weeks.
d. _____ You can see this frog's innards through translucent skin.
e. _____ Columbia frog could kill 10-20 humans.
f. _____ This frog's skin provides a healthy antibiotic.
g. _____ Grows back limb if it loses one.
h. _____ Produces its own antifreeze

SCRIPTURE

1. [The Lord] has made the earth _____, he has established the world _____, and has stretched out the heavens at His discretion. When He _____, there is a multitude of waters in the heavens: and He causes the vapors to ascend from the ends of the earth. He makes lightning for the rain, He brings the wind out _____." (Jeremiah 10)

2. And the LORD spoke to Moses, "Go to Pharaoh and say to him, 'Thus says the LORD: "Let My people go, that they may serve Me. But if you refuse to let them go, behold, I will smite all your territory with _____. So the river shall _____ _____, which shall go up and come into your house, into your _____, on your _____ into the houses of your servants, on your people, into your ovens, and into

GOD MADE ANIMALS - WORKBOOK

your _____. And the frogs shall come up on you, on your people, and on all your servants." (Exodus 8)

3. For behold, He who _____, and _____ _____, who declares to man what his thought is, and makes the morning darkness, who _____—The _____ _____ is His name. (Amos 4)

SPIRITUAL LIFE APPLICATION

1. How does God use frogs for His own purposes in the story of the Exodus? How does the Lord use frogs to help us today?

   _____
   _____
   _____
   _____
   _____
   _____
   _____
   _____

2. What are the most awesome designs discussed in this chapter that God has built into frogs and for which you would like to praise Him? Order the following according to their impressiveness to you (from 1 being most impressive down to 8). Take a moment to praise God for all of these qualities.

   Frog saliva _____
   Rapidity of tongue movement _____
   Waxy monkey tree frog pain killer _____
   Antifreeze in Alaskan frogs _____
   Metamorphosis of frog _____

Frog providing pregnancy test _____
Poison on the skin of poison frog _____
The flying frog's 50-foot flight _____

OBSERVATION OR DISSECTION

Choose one of the following exercises as you investigate God's world for yourself.

1. Dissect a frog using the guidelines contained on this web page: https://learning-center.homesciencetools.com/article/frog-dissection-project/

2. Observe a metamorphosis of a tadpole into a frog.

   a. Obtain frog eggs or tadpoles and watch the process of metamorphosis. Check state laws relating to removing frog eggs from public property. Check the temperature of the water where you have found the eggs or larvae. Take extra water out of the pond or lake where you collect the eggs or tadpoles. You will want to replicate this in your aquarium or bowl. Set up the habitat using the same water and try to keep the water at the same temperature as the pond where you found the eggs or tadpoles. DON'T USE CHLORINATED TAP WATER. You can let tap water sit for about a week to eliminate the chlorine. You can provide an air pump to provide oxygen for the water.

   b. Keep a log each day, either by describing in words or pictures what is happening. Your tadpoles will grow back legs first, and then front legs. At some point, the tadpole-turned-frog will stop breathing through the gills and start using his lungs instead.

   c. Feed the tadpoles algae or other small plant matter. You can grind up goldfish food real fine as well for a twice-a-day feeding. Or you could boil two tablespoons of lettuce or spinach, and then let it cool before feeding the tadpoles. You might add a few flakes of hard-boiled egg yolk once a week for a good protein.

   d. Once your tadpole begins to walk around, you will need to change his habitat around. Create a little slope out of the water, where the froglet can crawl out of the water. Use rocks and small branches.

3. Obtain a full grown frog or toad and observe its behavior. You might use a paper towel or gloves when handling the frog or toad.

a. Observe the frog or toad's skin. How is it different from your skin?

_____

_____

b. Release a fly in the aquarium or terrarium and observe. Take a video of the frog or toad capturing an insect with its tongue (if your phone/camera has that capability). Calculate the speed of the tongue. How long did it take to grab the insect?

_____

_____

_____

c. Measure the distance the frog or toad can hop. Compare that to its body length. How many body lengths can you jump?

_____

_____

_____

d. What is the frog's relaxed position?

_____

_____

e. Take a look at the frog's eyes. How do the eyes move? Move a pencil around the frog's body. Does it appear that the frog can follow this?

_____

_____

_____

CHAPTER 12 - OBSERVATION OR DISSECTION

Blow gently over the frog's eyes. What happens? Can you see the nictitating membrane (blinking eyelid)? Describe it.

_____

_____

_____

Shine a small light into the frog's eyes. What happens to the pupil? Does it dilate? With a parent's help, shine a light into your own eye. Does it do the same thing? Or is there a different response?

_____

_____

_____

f. Tilt the aquarium or terrarium back and forth. How does the frog respond? Does he lose his balance, or does he have some kind of balancing mechanism as humans do (in the inner ear)?

_____

_____

_____

g. Rub the sides of the frog and see if it will make a sound.

_____

_____

_____

h. Time the frog's breathing rate at room temperature. Watch his throat move up and down. Cool him off a little bit with an ice pack and then check his breathing rate again. Does it go up or down?

_____

_____

_____

i. Place the frog on its back and hold it somewhat firmly so it cannot move. What happens now? Why does the frog appear to be in a hypnotic state? What advantage does this give the frog?

Chapter 13
GOD MADE REPTILES

STUDY QUESTIONS

1. What are the three most deadly animals in the world (not including mosquitos)?

_____
_____
_____
_____

2. In what form did Satan appear when he tempted Eve and brought death into the world?

_____
_____

3. What would the serpent do to the Seed of the woman, according to the Lord's prophecy? What would the Seed of the woman (Jesus) do to the serpent?

_____
_____
_____

4. What did animals like dinosaurs and lions eat before the fall of Adam?

_____
_____
_____

GOD MADE ANIMALS - WORKBOOK

5. How big was the Titanosaur Patagotitan? How does that compare to a blue whale?

6. What are the indications that the behemoth in Job 40 is a dinosaur (and not an elephant or a hippo)?

7. How does a reptile differ from a bird?

8. How does a reptile differ from a fish?

9. What is the most important life-saving medical treatment provided for people bitten by a poisonous snake?

10. Fill in each blank with the animals from the list below that best match the description.

   Mary River turtle
   Tuatara
   King cobra
   Pit viper
   Reticulated python
   Belcher's sea snake

CHAPTER 13 - SCRIPTURE

Giant tortoise
Crocodile
Saw scaled viper

a. _____ Most venomous snake in the world
b. _____ Grows to 19 feet long
c. _____ Hibernates underwater
d. _____ Squeezes victims to death by restricting blood flow
e. _____ Snake with a heat sensor that takes in infrared signals
f. _____ Nocturnal reptile with third eye
g. _____ Deadliest snake in the world
h. _____ Lives for 200-250 years
i. _____ Sprays poison in victim's eyes

11. Fill in the correct number for the following snake characteristics:

   a. Snake's striking acceleration _____
   b. Largest king cobra _____
   c. Python's maximum squeeze force _____
   d. Largest Tyrannosaurus Rex tooth _____
   e. Number of people killed each year by snakes _____

SCRIPTURE

1. So the LORD God said to _____: "Because you have done this, you are cursed more than all cattle, and more than every beast of the field; _____ you shall go, and you shall eat dust all the days of your life. And I will put enmity between you and the woman, and between your seed and her Seed; He shall _____, and you shall _____."
(Genesis 3)

2. There are three things which are _____, Yes, four which I do not understand: The way of an _____, The way of an _____, The way of a ship in the midst of the sea, And the way of

GOD MADE ANIMALS - WORKBOOK

a man with a virgin. (Proverbs 30)

3. And He said to them, "I saw _____ fall like lightning from heaven. Behold, I give you the authority to _____, and over all the power of the enemy, and _____ _____. Nevertheless do not rejoice in this, that the spirits are subject to you, but rather rejoice because your names are written in heaven." (Luke 10)

SPIRITUAL LIFE APPLICATION

1. If Job knew about dinosaurs around 2000-1400 BC, when his book was written, why would this be a problem for modern evolutionists?

   _____
   _____
   _____
   _____
   _____

2. Read Luke 10:18-20 and Romans 16:20. Why should you not be afraid of snakes, and more specifically, Satan?

   _____
   _____
   _____
   _____
   _____
   _____
   _____

RESEARCH, OBSERVATION, AND DISSECTION

Choose one or more of the following exercises and complete the instructions.

1. Draw pictures of three different kinds of poisonous snakes. Identify the features on the snakes that would indicate that the snake was poisonous.

GOD MADE ANIMALS - WORKBOOK

2. Scientists have discovered soft skin tissue and proteins in dinosaur bones dug up in places from around the world. Research this website: https://answersingenesis.org/dinosaurs/bones/. Then explain how these discoveries disprove the idea that dinosaurs lived 80 million years ago.

_____
_____
_____
_____
_____
_____
_____
_____

3. Dissect a snake using guidelines included in the following websites: http://www.reptilesmagazine.com/Reptile-Care-For-Beginners/Snake-Anatomy/ or https://www.homesciencetools.com/content/reference/DE-GDSNAKE.pdf

Take a moment to praise God for the complexity, the functions, and the wisdom seen in the snake.

Describe the bone structure (or vertebrate) of the snake.

_____
_____
_____
_____

Describe the feel of the skin of the snake. Note how the scales on the underside of the body would be able to grip surfaces.

_____
_____
_____
_____

How many lungs do snakes have? How many kidneys?
___
___
___

How many chambers do we find in the snake's heart?
___
___
___

Carefully work the jawbones of the snake up and down. How flexible are the snake's jaws? How wide will the jaws open? What do you think could fit through these jaws?
___
___
___
___

Chapter 14

GOD MADE BOTH GREAT AND SMALL REPTILES

STUDY QUESTIONS

1. Where do we find the following animals?

   a. Komodo dragon _____
   b. Marine iguana _____
   c. Gila monster _____
   d. Chameleon _____
   e. Saltwater crocodile _____

2. What are the amazing characteristics God put in the gecko?
_____
_____
_____
_____

3. Which of the following animals can grow back part of their anatomy such as tails or legs (when severed)?

elephants	dogs	starfish
lizards	planarian flatworms	catfish
salamanders	penguins	grasshoppers

4. Which common trait do the shrew and the marine iguana share?
_____

GOD MADE ANIMALS - WORKBOOK

_____

_____

5. Fill in the animal which best fits each description below.

   a. _____ Most dangerous lizard in the world
   b. _____ A venomous reptile that can grow 10 feet in length
   c. _____ Filters salt out of its blood by an amazing system God designed for it.
   d. _____ Launches its tongue out to capture insects in a brief .07 seconds.
   e. _____ The force of this bite is six times the strength of a Rottweiler.
   f. _____ The force of this bite would crush a small car.

6. Where should you punch a crocodile if you ever find yourself in hand-to-claw combat with this animal?

_____

_____

_____

7. What are the differences between crocodiles and alligators?

_____

_____

_____

_____

8. What do crocodiles and alligators have in their mouths that could be dangerous to a person who is bitten by them?

_____

_____

_____

_____

CHAPTER 14 - SCRIPTURE

SCRIPTURE

1. He stretches out the north over empty space; He _____
_____. He binds up the water in His thick clouds, yet the clouds are not broken under it. He covers the face of His throne, And spreads His cloud over it. He drew a circular horizon on the face of the waters, at the boundary of light and darkness. The pillars of heaven tremble, and are astonished at His rebuke. He _____ with His power, and by His understanding _____. By His Spirit He adorned the heavens; His hand _____. Indeed these are the mere edges of His ways, and how small a whisper we hear of Him! But _____ who can understand? (Job 26)

2. Benaiah was the son of Jehoiada, the _____, who had done many deeds. He had killed two lion-like heroes of Moab. He also had gone down and ___ _____ on a snowy day. (2 Samuel 23)

3. O LORD, our Lord, How _____, Who have set _____! (Psalm 8)

SPIRITUAL LIFE APPLICATION

1. Read Revelation 12:11. How do we overcome the powerful dragon (Satan) in our spiritual warfare? List the three strategies for our victory. Write out your testimony of faith in Jesus in the space below.

_____
_____
_____
_____
_____

GOD MADE ANIMALS - WORKBOOK

2. Study Job 26:7-14 and answer the following questions:

   a. How is God's power seen in creation?

   b. How is God's power seen over the serpent?

RESEARCH, OBSERVATION, AND DISSECTION

Choose one or more of the following exercises and complete the instructions.

1. Obtain a live lizard, gecko, or chameleon in your garden or from a friend who keeps a lizard as a pet. Examine the lizard and answer the following questions:

   a. What kind (or species) of lizard is it?

b. Examine the lizard's skin? Is it wet or dry?

c. Put your lizard in water. Observe how it swims. How often does it take a breath of air? Does it hold its head outside of the water?

d. How energetic is your lizard? Cool off your lizard a little bit. Does it become less energetic?

e. What does this lizard eat?

f. How long is the lizard's tongue?

g. Is there anything unusual about your lizard? Does he do push-ups? Does he bob his head? Does he wave his tail? Does he change his colors?

2. Experiment with static electricity. This is not the same thing as the gecko's vanderwaal force, but it is somewhat similar.

   Blow up a balloon and rub it on a sweater or your head for about 30 seconds. Stick it on the ceiling. How long does it take before the balloon falls off the ceiling?

   _____
   _____
   _____
   _____

   By rubbing two different materials together, a bunch of electrons will gather on the surface of the material (the balloon), creating a negative electric charge. When you rub the materials together, you are increasing the surface area contact between the two materials and the electric charge builds up faster. Because opposites repel each other, the electrons sitting on the surface of the ceiling pull away from the balloon, leaving a positively charged surface. The positive charge attracts the negative charge (the balloon), and the balloon sticks for a while.

Chapter 15
GOD MADE MAMMALS

STUDY QUESTIONS

1. What are the distinguishing features of the following animals?

   a. Mammals
   _____
   _____
   _____

   b. Amphibians
   _____
   _____
   _____

   c. Reptiles
   _____
   _____
   _____

   d. Birds
   _____
   _____
   _____

GOD MADE ANIMALS - WORKBOOK

e. Exoskeleton vertebrates
_____
_____
_____

2. Provide examples of the following types of mammals.

   a. Pouched Mammals _____
   b. Placenta Mammals _____
   c. Egg-laying Mammals _____
   d. Insect-eating Mammals _____
   e. Flesh eating Mammals _____
   f. Gnawing Mammals _____
   g. Sea Mammals _____
   h. Winged Mammals_____
   i. Proboscis Mammals_____
   j. Even-toed Mammals _____
   k. Odd-toed Mammals _____
   l. Semi-erect Mammals_____

3. How do the following warm-blooded mammals/humans control their temperatures?

   a. Rats
   _____
   _____

   b. Dogs
   _____
   _____

   c. Humans
   _____
   _____

4. What are the most intelligent mammals?
   _____

CHAPTER 15 - STUDY QUESTIONS

5. What are the animals that become the most attached to humans?

6. How do we know that the Neanderthals were human?

7. What are some of the most amazing defense mechanisms the Lord has created in the following mammals?

   a. Three-banded armadillo _____
   b. Dormouse _____
   c. Porcupine _____
   d. Hedgehog _____
   e. Opossum _____

8. Match the ability to the animal God created.

Cockroaches	Fashions its own house with rooms and openings
Parrots	Talks by mimicry
Crows	Walks on two feet for short distances
Beavers	Makes tools to use

9. List the animals that are described by each quality:

   a. Majestic Animals _____
   b. Largest Land Animal _____
   c. Dangerous Mammals _____

GOD MADE ANIMALS - WORKBOOK

d. Unusual Mammal _____
e. Beautiful Animals _____
f. Funny-looking Animals _____
g. Cutest Animals _____

10. Provide definitions for the following terms:

    a. Placenta_____
    b. Belly button _____
    c. Marsupial_____
    d. Nocturnal _____
    e. Diurnal_____
    f. Archeologists _____

SCRIPTURE

1. Save me from the _____ And from the _____!
   (Psalm 22)

2. There are three things which are _____, Yes, four which are _____. _____among beasts, And does not turn away from any. (Proverbs 30)

3. All flesh is not the _____, but there is one kind of flesh of men, another flesh of _____, another of _____, and another of _____. (1 Corinthians 15)

4. One generation shall praise Your works to another, and shall declare Your _____. I will meditate on the glorious splendor of Your majesty, And on Your _____. Men shall speak of the might of Your _____, And I will declare Your greatness. They shall utter the ___

136

_____, and shall sing of Your righteousness. The LORD is gracious and full of compassion, slow to anger and great in mercy. The LORD is good to all, and His tender mercies are over _____. _____ shall praise You, O LORD, and Your saints shall bless You. They shall speak of the glory of Your kingdom, and talk of Your power... (Psalm 145)

SPIRITUAL LIFE APPLICATION

1. What are the functions that we share with mammalian animals? What are some human functions that the mammalian animals do not share with us? How has God made you with more dignity than the animals? Take a moment to thank the Lord for providing us with dignity and honor. Thank the Lord also for the blessings of the mammalian traits He has given to us.

_____
_____
_____
_____
_____
_____
_____
_____
_____
_____

2. Of all of God's created mammals, which do you find the most impressive? Why? Give God the glory for these amazing creatures!

_____
_____

GOD MADE ANIMALS - WORKBOOK

3. Identify several ways in which you can see the kindness of God in respect to mammals. Give thanks for God's kindness to you as well.

OBSERVATION OR DISSECTION

Choose one of the following exercises as you investigate God's world for yourself.

1. Study a mammal, preferably a dog. The questions below may be adapted for a cat.

    a. Is this dog a male or female? How can you tell the difference?

    b. Examine the dog's hair for healthy/unhealthy conditions. A healthy coat of hair would be smooth and shiny, not course and brittle. You are looking for healthy skin--not greasy, bumpy, or flaky.

c. Examine the dog's nose. Healthy noses are usually moist and cool. The dog's nose is a very important sensory device provided by the Creator—operating at 1000-10,000 times the sensitivity of a human nose. The nose skin contains glands which secrete a mucus that keeps the dog's smelling finely tuned. Dogs rely on their noses more than their eyesight. Also, the nose is another means of keeping the dog cool on hot days.

_____
_____
_____

d. Check the dog's feet for foxtails between the toes or ruptured blisters. Check the dog's toenails. Generally, the nails should be trimmed monthly so the tip of the nail is pretty much even with the pad on the dog's toe.

_____
_____
_____

e. What does the dog eat? How much does it eat?

_____
_____
_____

f. What is its personality type? Is it confident or submissive? Is it easily frightened? Is it friendly? How active is the dog? How sociable is the dog?

_____
_____
_____
_____

g. How much does the dog guard its toys or food? How does the dog mark its territory?

_____
_____
_____

GOD MADE ANIMALS - WORKBOOK

h. What are the types of noises or stimuli that encourage the dog to be afraid?

_____

_____

_____

_____

2. Observe nocturnal mammals and animals sometime after dark. Forested areas can be teeming with life at night. Shine a flashlight about, and look for the glow of eyes in the dark. This is called spotlighting. You might cover the flashlight with red cellophane so as not to harm the animal's eyes. Look for mammals, reptiles, amphibians, birds, snakes, and insects. Make notes of what you see. Research for the animal names. Try to identify what they eat and where they like to hunt for food.

_____

_____

_____

_____

_____

_____

_____

_____

_____

3. Order a cow eye from a butcher shop, and proceed with a dissection using the instructions located here: https://www.exploratorium.edu/learning_studio/cow_eye/coweye.pdf

The cow's eye is somewhat like a human eye, amazing designs provided by our Creator God. It is far more complex and sophisticated than a camera lens. The eye quickly adjusts to various levels of light all by itself. It can focus on something close in and something far away in a split second. The entire image of what it sees is passed on to the brain, so we can immediately know what is happening around us.

Praise God for the amazing instrument He created for cows, and for us!

Chapter 16
TAKING DOMINION OF THE MAMMAL WORLD

STUDY QUESTIONS

1. What animals were create with the capability of sonar and infrared detection systems?

2. What was it that motivated man to create sonar?

3. What are the blessings that fruit bats bring to us?

4. What are the first and second most common meats in the world? How much meat is produced from these animals each year?

GOD MADE ANIMALS - WORKBOOK

5. How much wasted land is there in Africa?

6. What are the benefits and the drawbacks of the following animals (if you are considering raising them)?

   a. Cows

   b. Sheep

   c. Goats

   d. Pigs

7. How are the following animals helpful to man?

   a. Asian elephant

   b. Pangolin

CHAPTER 16 - STUDY QUESTIONS

c. Camel

d. Pigeons

e. Elk

f. Mongoose

g. Lizard

h. Angora rabbit

8. Mark the following with "Good Dominion" if it is an example of humans taking effective rule over animals, or "Bad Dominion" if it is an example of humans taking ineffective rule over animals.

a. _____ Releasing wolves into Idaho and Wyoming in 1995.
b. _____ Refusing to pen up an animal that is known to bite humans and other animals.
c. _____ Training a dog to lay in his crate without whining or barking.
d. _____ Raising buffalo on private ranches
e. _____ Disallowing logging on public lands in the later 1980s in America.

GOD MADE ANIMALS - WORKBOOK

9. What dog breed would be best for the following purposes:

    a. Pulling sleds in the snow _____
    b. Herding sheep _____
    c. Hunting dog _____
    d. Guard dog_____

SCRIPTURE

1. Then the LORD sent Nathan to David. And he came to him, and said to him: "There were two men in one city, one rich and the other poor. The rich man had exceedingly _____ _____. But the poor man had nothing, except one _____ which he had bought and _____; and it grew up together with him and with his children. It ate of his own food and drank from his own cup and _____; and it was like _____ _____. (2 Samuel 12)

2. Now while he was still speaking with them, Rachel came with _____ _____, for she was _____. And it came to pass, when Jacob saw Rachel the daughter of Laban his mother's brother, and the sheep of Laban his mother's brother, that Jacob went near and rolled the stone from the well's mouth, and _____ _____his mother's brother. (Genesis 29)

3. If an ox_____, then the ox shall surely be stoned, and its flesh shall not be eaten; but the owner of the ox shall be acquitted. But if the ox tended to _____, and it has been made known to his owner, and he has not kept it confined, so that _____ _____, the ox shall be_____ and its owner also shall _____. (Exodus 21)

146

CHAPTER 16 - SPIRITUAL LIFE APPLICATION

4. Be diligent to know _____, and attend to your _____; for riches are not forever, nor does a crown endure to all generations. When the hay is removed, and the tender grass shows itself, and the herbs of the mountains are gathered in, the lambs will _____, and the goats _____; you shall have enough _____ _____, for the _____, and the nourishment of your maidservants. (Proverbs 27)

SPIRITUAL LIFE APPLICATION

1. Read Genesis 1:26 and Proverbs 27:23-27. Why should you take care of God's World? What are the rules that govern our treatment of the animals and land?

   _____
   _____
   _____
   _____
   _____

2. What is your favorite dog rescue story? What is the best blessing that dogs have brought to the world? Take a moment to thank God for the blessing of dogs.

   _____
   _____
   _____
   _____

3. How much beef does your family eat in a given year? How much milk does your family consume in a year? Given that number, try to calculate how many cows are consumed in your town in a given year. Take a moment to thank the Lord for all of that food He provides through cows.

   _____
   _____

GOD MADE ANIMALS - WORKBOOK

_____
_____
_____
_____

OBSERVATION OR DISSECTION

Choose one of the following exercises as you investigate God's world for yourself.

1. Study the best ways to protect properties from forest fires and protect your trees from bug kill. What are the bugs that tend to kill trees, and what is the best way to protect your trees?

   _____
   _____
   _____
   _____
   _____
   _____
   _____
   _____
   _____
   _____

2. Examine the following pictures of ten animals. Name the animal and its class (example: "horse, mammal").

A. CLASSIFICATION:

_____
_____
_____
_____
_____
_____

CHAPTER 16 - OBSERVATION OR DISSECTION

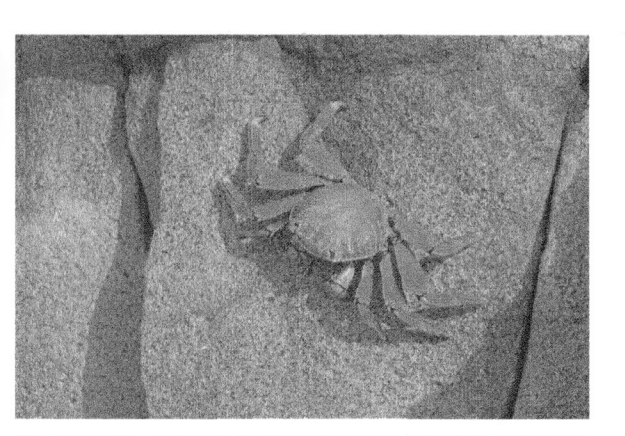

B. CLASSIFICATION:
_____
_____
_____
_____
_____
_____
_____

C. CLASSIFICATION:
_____
_____
_____
_____
_____
_____
_____

D. CLASSIFICATION:
_____
_____
_____
_____
_____
_____
_____

E. CLASSIFICATION:
_____
_____
_____
_____
_____
_____
_____

149

GOD MADE ANIMALS - WORKBOOK

F. CLASSIFICATION:
_____
_____
_____
_____
_____
_____

G. CLASSIFICATION:
_____
_____
_____
_____
_____
_____

H. CLASSIFICATION:
_____
_____
_____
_____
_____
_____

I. CLASSIFICATION:
_____
_____
_____
_____
_____
_____

CHAPTER 16 - OBSERVATION OR DISSECTION

J. CLASSIFICATION:

_____
_____
_____
_____
_____
_____
_____

3. Visit a farm or home where a dog, cat, goat, or sheep has just had a baby. Observe how the mother takes care of her babies. Watch the babies nurse. Make notes below of how the mother dog/cat/goat/sheep care for their young. You may be able to feed a baby lamb or a baby goat with a bottle.

_____
_____
_____
_____
_____
_____
_____
_____
_____

Or visit a farm where the sheep are being sheared. See if you can spin the fiber into yarn using a spinning wheel.

FINAL EXAMINATION

30 POINTS TOTAL

A. TRUE OR FALSE

1. _____A beetle has an exoskeleton.
2. _____Cephalopods have their head in their foot.
3. _____No mammals lay eggs.
4. _____All arthropods are insects.
5. _____Arachnids have eight legs.
6. _____Frogs and butterflies go through metamorphosis.
7. _____Moths are different from butterflies in that they have a club at the end of their antennae.
8. _____Sharks are different from fish in that their skeletons are made of cartilage.
9. _____The tick is an arachnid.
10. _____The largest mammal in the world is the elephant.
11. _____The wine Jesus made at Cana had the appearance of age.
12. _____Birds do not sweat because they are cold-blooded.
13. _____Insects do not bleed.
14. _____The most dangerous animal in the world, causing the most deaths to humans, is the mosquito.
15. _____The kangaroo feeds its young by a placenta.

GOD MADE ANIMALS - WORKBOOK

B. DEFENSE MECHANISMS

What sort of defense mechanism has God given the following animals? Name at least one unique defense mechanism.

16. Opossum _____
17. Clown fish _____
18. Lion fish _____
19. Three-banded armadillo _____
20. Chameleon _____
21. Loris _____

C. UNIQUE GIFTS

Name the unique gifts described below that the Creator has provided for the animals.

22. The ability of bats to fly at night and locate objects.
_____
_____

23. Certain pit vipers have this ability to identify heat sources.
_____
_____

24. Mammals have these glands to feed their babies.
_____
_____

25. The lines on fish which are able to detect pressure changes in the water.
_____
_____

FINAL EXAMINATION

D. SHORT ANSWER

26. What two chapters in the Bible instruct us to take dominion of the animals?
_____
_____

27. What treatment do doctors use on people who have been bitten by a poisonous snake or spider?
_____
_____

28. What have researchers found in dinosaur bones that indicate the animals were buried much more recently than 80 million years ago?
_____
_____

29. Animals without backbones are called _____. An example would be _____.

30. Animals used to clean up dead bodies are called _____. An example would be _____.

GOD MADE ANIMALS - WORKBOOK

CHAPTER 1

STUDY QUESTIONS

1. Fleming
2. Kelvin
3. Newton
4. Boyle
5. Stokes
6. Babbage
7. Pasteur
8. Faraday
9. God
10. Chance/accidents
11. God's Power and Strength
12. God's Wisdom and Knowledge
13. God's Goodness
14. God
15. They believe Grandfather exists. They believe Grandfather is good and he tells the truth. They believe Grandfather has not made it too difficult to find the treasures. They believe Grandfather does not change the location of the treasures. They believe Grandfather has told them to look for the treasures.
16. God promised Noah He would not curse the ground and destroy all creatures again as He had in the flood. He promised Noah that regular seasons and patterns would continue without interruption.
17. Unbelieving scientists don't sing and they don't praise God because they don't believe that a personal God has created all of this wonderful creation.
18. We study science to glorify God and praise Him for His amazing works. We study science so we can take dominion and rule over the creation. God has told us to do this, and we must obey God. We study science to know God.

SCRIPTURE

1. "The heavens declare the **glory of God**; and the firmament shows **His handiwork**. Day unto day utters speech, and night unto night reveals knowledge. There is no speech nor language **where their voice is not heard.**" (Psalm 19:1-3)

2. "So God created man in His own image; in the image of God He created him; male and female He created them. Then God blessed them, and God said to them, 'Be **fruitful and multiply**; fill the earth and subdue it; have dominion over **the fish of the sea**, over **the birds of the air**, and over every living thing that **moves on the earth.**'" (Genesis 1:27-28)

3. "Then the LORD said in His heart, 'I will never again curse the ground for man's sake, although the imagination of man's heart is evil from his youth; nor will I again destroy every living thing as I have done. While the earth remains, **seedtime and harvest**, cold and heat, winter and summer, and **day and night shall not cease.**'" (Genesis 8:21-22)

4. "It is the glory of God to **conceal a matter**, but the glory of kings **is to search out a matter.**" (Proverbs 25:2)

CHAPTER 2

STUDY QUESTIONS

1. Creation Day 1 (Genesis 1:1-5) — God created the heavens and the earth. The heavens mean space outside of the material earth. The earth is matter. And, God created light.

2. Creation Day 2 (Genesis 1:6-8) — God created the sky with clouds and moisture in the air.

3. Creation Day 3 (Genesis 1:9-13) — God created dry land and the oceans. God made all plant life.

4. Creation Day 4 (Genesis 1:14-19) — God created the moon, the sun, and the stars in the sky.

5. Creation Day 5 (Genesis 1:20-23) — God made fish and birds. He also made flying insects, and he gave these animals the ability to reproduce (or have babies).

6. Creation Day 6 (Genesis 1:24-31) — God created all the animals that live on dry land. And finally, God made man in His own image.

7. Unbelieving evolutionists believe these things came up by natural processes and by accident (not by a personal, intelligent Creator).

8. Tables — wood (usually)
Roads and highways — rocks, sand, and limestone.
Car hoods — iron ore
Tires — rubber (rubber trees)

9. God made the world and all the animals out of nothing.

10. Jesus made wine out of water.

11. Jesus made all things (Ps. 33:6), He raised Lazarus from the dead (John 11:44), He calmed the seas when He said, "Peace, be still" (Mark 4:39), and He upholds all things by the word of His power (Heb. 1:3).

12. Mankind. God created man out of the dust of the earth.

13. He created man as a grown up. He created trees with rings already in place. He created the stars in the sky, with the appearance that they had been there for a long time.

14. The chicken.

15. Our Lord Jesus used five barley loaves and two small fish. He multiplied the fish and loaves 5,000 times, and added matter to the world, where the matter had not existed before.

SCRIPTURE

1. Let them praise the name of the LORD: for **He commanded**, and they were created. (Psalm 148:5)

2. "By the **word of the LORD** the heavens were made, and all the host of them by **the breath of His mouth**." (Psalm 33:6)

3. "Then God said, 'Let the waters abound with **an abundance of living creatures**, and let birds fly above the earth across the face of the firmament of the heavens.' So God created great sea creatures and every living thing that moves, **with which the waters abounded**, according to their kind, and every winged bird according to its kind. And God saw that it was good." (Genesis 1:20-21)

CHAPTER 3

STUDY QUESTIONS

1. Answers may vary. The real dog can reproduce. The real dog has a unique personality. It can run much faster. It can eat. It is life that God has created.

2. They cannot breed. For example a cat and dog cannot breed to produce a cat-dog.

3. Answers may vary.
a. Category Monera -- strepococcus and staphylococcus
b. Category Protista -- Algae and Protozoans (like amoebas)
c. Category Fungi -- mushrooms and molds
d. Category Plantae -- trees, plants, flowers, etc.
e. Category Animalia -- insects, birds, fish, mammals, etc.
f. Category Man -- You and me

4. Man is made in the image of God. Man is capable of relationship with God. He is created with a sense of righteousness and morality. He can know a million times what an animals knows. Man can

communicate on a very complex level using many forms of communication. Man is also capable of emotion. He is capable of appreciating God, and appreciating God's creation. He can appreciate art. He creates art. Animals cannot do any of this.

5. Man has two legs, some birds have two legs. Man has lungs and a heart, as do mammals. Man has a brain, as do some animals. Man has ears, eyes, a nose, and a mouth, as do some animals. Man can walk and run, as do some animals.

6. a. Species
   b. Cytoplasm
   c. Cell
   d. Genetic code (DNA)
   e. Reproductive system

7. a. Zonkey
   b. Coydog
   c. Cannot breed
   d. Cockapoo
   e. Wolfdog
   f. Mule
   g. Cannot breed

8. No animals ate meat before the fall.

9. Answers may vary. Lions, wolves, dogs, bears, dinosaurs, etc.

10. After the worldwide flood (at the time of Noah).

SCRIPTURE

1. "Then God said, 'Let Us make man in **Our image**, according to **Our likeness**; let them have dominion over the fish of the sea, over the birds of the air, and over the cattle, over all the earth and over every creeping thing that creeps on the earth.' So God created man **in His own image**; **in the image of God** He created him; male and female He created them." (Genesis 1:26-27)

2. "What is **man** that You are mindful of him, And the **son of man** that You visit him? For You have made him a little lower than the angels, And You have crowned him with glory and honor. You have **made him to have dominion over the works of Your hands**; You have put **all things under his feet**…" (Psalm 8:4-6)

3. "Out of the ground the LORD God formed every beast of the field and every bird of the air, and brought them to Adam to see what **he would call them**. And whatever Adam called each living creature, that was its **name**. So Adam gave **names to all cattle**, to the birds of the air, and to every beast of the field. But for Adam there was not found a **helper comparable** to him." (Genesis 2:19-20)

4. "For we know that the whole creation **groans and labors with birth pangs** together until now. Not only that, but we also who have the firstfruits of the Spirit, even we ourselves groan within ourselves, eagerly waiting for the adoption, **the redemption of our body**." (Romans 8:20-23)

SPIRITUAL LIFE APPLICATION

1. God => Angels => Man => Animals. We rule over the sheep, the oxen, the fish, and the birds.

CHAPTER 4

STUDY QUESTIONS

1. a. Cephalopod
   b. Coral
   c. Invertebrate
   d. Gastropod
   e. Arthropod
   f. Bivalve mollusk
   g. Metamorphosis
   h. Arachnid
   i. Crustacean
   j. Insect

2. Sydney funnel-web spider. The Brazilian wandering spider is the most venomous (but it is not as much a

ANSWER KEY

threat to people as the Sydney funnel-web).

3. Ticks suck the blood of animals and humans, and diseases are easily transmitted through blood.

4. A chigger is a baby mite (in the larvae stage). It burrows under the skin and lives on blood.

5. a. Crickets, Grasshoppers, Locusts, Cockroaches, Beetles
True bugs like bedbugs, water striders, and stinkbugs
Aphids, leaf hoppers, cicadas, and lac insects
Dragonflies and damselflies
Social insects like ants, bees, and wasps
Butterflies and moths
Flies, gnats, and mosquitoes
b. Clams, oysters, scallops
c. Cowries, conches
d. Snails, slugs
e. Squid, octopus
f. Spiders, black widow, tarantula, etc.
g. Butterflies, moths, frogs
h. Shrimp, lobsters, crabs

6. Octopus -- Can poison victims, escape by jet propulsion, and spray black fluid to confuse its predators.
Mollusks -- God gave them hard shells in which they can hide.
Jellyfish -- God gave them stingers on their tentacles.
Flea -- God gave it the ability to jump 150 times its own height.
Bombardier beetle -- God gave it the uncanny ability to fire an explosion out its rear end, intended to fry any would-be attackers.
Spiders -- God gave many of them strong toxins to ward off predators.

7. The Ant.

8. The moth's antennae is feathery and fuzzy, while the butterfly's is smoother with a fat knob on the end of it. The butterfly likes to fold her wings up, while the moth looks a little more depressed. She folds her wings down over her abdomen. Also, moths tend to be smaller and less colorful.

9. a. Egg
b. Caterpillar
c. Chrysalis
d. Butterfly

10. Hard work and teamwork.

11. a. Earthworms are great for farms because they keep the ground broken up and porous enough to help soak the rainwater into the dirt.
b. Crustaceans are God's cleanup crew to clean up the dead animals at the bottom of the ocean.
c. Spiders eat bad insects like mosquitos, flies, etc.

12. Ticks, poisonous and dangerous spiders, slugs, and snails.

13. a. Trap door spider
b. Goliath birdeater
c. Water spider, or Diving bell spider
d. Ogre-faced spider
e. Darwin's bark spider

SCRIPTURE

1. 1. Praise the LORD from the earth,
You great sea creatures and all the depths;
Fire and hail, snow and clouds;
Stormy wind, fulfilling His word;
Mountains and all hills;
Fruitful trees and all cedars;
Beasts and all cattle;
Creeping things and flying fowl.
(Psalm 148:7-10)

2. The LORD is gracious and full of compassion,
Slow to anger and great in mercy.
The LORD is good to **all**,
And His tender mercies are over **all His works**.
(Psalm 145:8-9)

3. This great and wide **sea**,
In which are **innumerable teeming things**,
Living things both small and great.
There the ships sail about;
There is that **Leviathan**
Which You have made to play there.
These all wait for You,

GOD MADE ANIMALS - WORKBOOK

That You may give them **their food in due season**.
(Psalm 104:24-27)

4. There are. . .things which are **little on the earth**,
   But they are exceedingly wise:
   The **ants are a people** not strong,
   Yet they prepare their food in the summer.
   (Proverbs 30:24-25

5. There be. . . things which are little upon the earth,
   but they are exceeding wise:
   The **spider taketh hold with her hands**, and is in kings' palaces. (Proverbs 30:24, 28 KJV)

CHAPTER 5

STUDY QUESTIONS

1. a. Crustaceans (lobsters, shrimp, etc.)
   b. Earthworms
   c. Dung Beetles

2. Insectivores

3. Carnivores

4. Herbivores

5. Omnivores

6. The mosquito can spread diseases like Malaria, Yellow Fever, Zika Virus, and Dengue Fever, killing 250,000 people a year.

7. The fly can see 360 degrees around, and picks up movement five times faster than we can.

8. The fly passes diseases around by crawling around disgusting piles of garbage and dead animals and passing the pathogens on to human food and surfaces which humans touch.

9. Lyme disease, Typhus, Bubonic plague

10. Aphids (also white flies, locusts, stink bugs, Colorado potato beetle, the Khapra beetle, and the Japanese beetle)

11. Ladybugs, praying mantises

12. The pollen is the stamen of the plant.

13. The female cell is at the very center of the carpel.

14. Bees brush up against the stamen and get pollen on their hairy little bodies, and then the powder falls off of their bodies into the carpel of other plants (where the pollen contacts the female cells or the ovule).

15. Anticoagulant, Anesthetic, Vasodilator, Antibiotic

SCRIPTURE

1. Then God spoke to Noah, saying, "Go out of the ark, you and your wife, and your sons and your sons' wives with you. Bring out with you every living thing of all flesh that is with you: birds and cattle and every **creeping thing that creeps on the earth**, so that they may **abound** on the earth, and be **fruitful and multiply** on the earth." (Genesis 8:15-17)

2. The fear of the LORD is clean, enduring forever; the **judgments of the LORD** are true and righteous altogether. More to be desired are they than gold, yea, than much fine gold; sweeter also than **honey and the honeycomb**. Moreover by them Your servant is warned, and in keeping them there is great reward. (Psalm 19:9-11)

3. He causes the grass to grow for the cattle,
   And vegetation for **the service of man**,
   That he may bring forth food from the earth,
   And **wine that makes glad the heart of man**,
   Oil to make his face shine,
   And **bread** which strengthens man's heart.
   The trees of the LORD are full of sap,
   The cedars of Lebanon which He planted,
   Where the birds make their nests;
   The stork has her home in the fir trees.
   (Psalm 104:14-17)

ANSWER KEY

CHAPTER 6

STUDY QUESTIONS

1. Land, sea, and air
2. Penguins, skua bird, polar bears, Arctic foxes, ringed seals
3. The bar headed goose and the Himalayan jumping spider
4. Leviticus 17:10-12, Acts 15:28-29
5. Flat worms, coral, and jellyfish are bloodless. Frogs, birds, fish, crabs, lobsters, starfish, mammals, and man have life in the blood.
6. Warm-blooded animals can regulate their blood and body temperature. Cold-blooded animals cannot internally control their blood and body temperature.
7. Warm-blooded animals include mammals and birds. Cold-blooded animals include snakes and other reptiles, frogs, butterflies, earthworms, and most fish.
8. Dogs and birds will pant to cool themselves off. The human body cools itself off by sweating.
9. Some birds and mammals.
10. First, the bird swallows the food into the stomach where hydrochloric acid breaks down the food. Then, the food moves down into the gizzard where it is ground up using small stones which the bird has swallowed for that purpose.
11. a. Arctic tern
    b. Cormorant
    c. Common poorwill
    d. Bar headed goose

SCRIPTURE

1. Every moving thing that lives shall be food for you. I have given you all things, even as the green herbs. But you shall not eat **flesh with its life**, that is, its **blood**. (Genesis 9:3)

2. For He says to the **snow**, "Fall on the earth;"
   Likewise to the gentle rain and the heavy rain of His strength.
   He seals the hand of every man,
   That all men may know His work.
   The **beasts go into dens**,
   And remain in their **lairs**.
   (Job 37:6-8)

3. But those who wait on the **LORD**
   Shall renew their strength;
   They shall mount up with **wings like eagles**,
   They shall run and not be **weary**,
   They shall walk and not **faint**.
   (Isaiah 40:31)

4. Therefore I say to you, do not worry about your life, **what you will eat or what you will drink**; nor about your body, what you will put on. Is not life more than food and the body more than clothing? Look at **the birds of the air**, for they neither sow nor reap nor gather into barns; yet your heavenly Father feeds them. Are you not of more value than they? (Matthew 6:25-27)

SPIRITUAL LIFE APPLICATION

1. God feeds about 3.6 trillion birds, fish, and mammals. Humans feed about 15 billion animals. The ratio is 240x. If you include insects, God feeds 33,000,000 times more animals than humans do. You should not worry about your material needs because, if God feeds all those animals, He can feed you.

CHAPTER 7

STUDY QUESTIONS

1. They studied the birds.

GOD MADE ANIMALS - WORKBOOK

2. Light bones, top of wing designed with slight hump, adjustment of wing tips independently, the alula bone, and flexible adjustable feathers for maximum dexterity in flight.

3. a. Chimney swift
   b. Golden eagle
   c. Parrot
   d. New Caledonian crow
   e. African hamerkops
   f. Woodpecker
   g. Peregrine falcon

4. January-March -- Adults feed
   April-May -- Mating at the rookery
   June-July -- Females leave and males sit on eggs
   August -- Eggs hatch
   September-October -- Adults take turns feeding chicks
   October-December -- Chicks learn to huddle together for warmth and feed themselves

5. a. Satin bower bird -- Decorates nesting area with blue objects
   b. Hummingbird -- Dives straight down at 60 mph, producing a loud whistle
   c. Great crested grebes -- Performs graceful duet dances
   d. Puffin -- Sports bright colored beaks

6. Laysan albatross, mute swan, bald eagle, black vulture, scarlet macaw, whooping crane, and Atlantic puffin.

SCRIPTURE

1. He sends **the springs into the valleys**;
   They flow among the hills.
   They give drink to every beast of the field;
   The wild donkeys quench their thirst.
   By them **the birds of the heavens have their home**;
   They **sing** among the branches.
   He waters the hills from His upper chambers;
   The earth is satisfied with the fruit of Your works.
   (Psalm 104:10-13)

2. But now ask the beasts, and they will teach you;
   And the **birds of the air**, and they will tell you;
   Or speak to the earth, and it will teach you;
   And the fish of the sea will explain to you.
   Who among all these does not know
   That **the hand of the Lord** has done this,
   In whose hand is the **life of every living thing**,
   And the breath of all mankind?
   (Job 12:7-10)

3. Even the **sparrow** has found a home,
   And the swallow **a nest for herself**,
   Where she may lay her young—
   Even Your   , O LORD of hosts,
   My King and my God.
   Blessed are those who **dwell in Your house**;
   They will still be praising You.
   (Psalm 84:3-4)

CHAPTER 8

STUDY QUESTIONS

1. a. Helpful
   b. Helpful
   c. Helpful
   d. Harmful
   e. Helpful
   f. Helpful
   g. Harmful
   h. Harmful
   i. Harmful
   j. Helpful
   k. Harmful

2. a. Raccoons
   b. Dead rabbits
   c. Dead rodents/fleas

3. Domesticated birds include chicken, geese, and ducks. Domesticated mammals include dogs, cats, cows, goats, sheep, etc.

164

4. Pigeons are used to carry messages, and they make good meat birds.

5. Chickens provide meat, eggs, and good fertilizer.

SCRIPTURE

1. Our soul has escaped as a **bird from the snare** of the fowlers;
The snare is broken, and we have escaped.
Our help is in **the name of the LORD**,
Who made heaven and earth.
(Psalm 124:7-8)

2. Be diligent to know the **state of your flocks**. And attend to your **herds**; for riches are not forever, nor does a crown endure **to all generations**.
(Proverbs 27:23-24)

3. Now Abel was a **keeper of sheep**, but Cain was a tiller of the ground. And in the process of time it came to pass that Cain brought an offering of the fruit of the ground to the LORD. Abel also brought of the **firstborn of his flock** and of their fat. And **the LORD respected Abel and his offering**.
(Genesis 4:2-4)

CHAPTER 9

STUDY QUESTIONS

1. a. Vertebrates
   b. Invertebrates
   c. Vertebral column
   d. Skull
   e. Water
   f. Cartilage

2. Gills

3. The pectoral, the pelvic, the dorsal, the anal, and the caudal fins

4. The caudal fin (or the tail fin)

5. a. Marlin
   b. Spotted ray
   c. Stonefish
   d. Black swallower
   e. Archer fish
   d. Mandarin fish
   e. Walking catfish
   f. Hawaii goby fish

6. Boxer crab -- sea anemone
   Blue streak wrasses -- oriental sweetlips
   Imperial shrimp -- sea slug
   Goby fish -- blind pistol shrimp

7. a. Porcupine fish — Spines that look intimidating and an orange coloring.
   b. Surgeon fish -- A blade that whips out like a switchblade to slice up a predator
   c. Stingray -- A stinger
   d. Lion fish -- Spiny rays that can shoot venom into predators
   e. Coral reef fish -- Coloring that blends in perfectly with the coral
   f. Mandarin fish -- Poisonous and smelly

SCRIPTURE

1. Oh, give thanks to the LORD, for **He is good**!
For His mercy endures forever. . .
Who **gives food to** all flesh,
For His mercy endures forever. (Psalm 136:1,25)

2. Then God said, "Let the waters abound **with an abundance of living creatures**, and let birds fly above the earth across the face of the firmament of the heavens." So **God created great sea creatures** and every living thing that moves, with which the waters abounded, according to their kind. . . (Genesis 1:20-21a)

3. For the creation **was subjected to futility**, not willingly, but because of Him who subjected it in hope; because the creation itself also will be delivered from the bondage of corruption into the glorious liberty of the children of God. For we know that **the whole creation groans and labors** with birth pangs together until now. Not only that, but we also who have the firstfruits of the Spirit, even we

ourselves groan within ourselves, eagerly waiting for the adoption, the **redemption of our body**. (Romans 8:20-23)

able to destroy both soul and body in hell. Are not two sparrows sold for a copper coin? And not one of them falls to the ground apart **from your Father's will**. But the very hairs of your head are all numbered. Do not **fear** therefore; you are of more value than many sparrows. (Matthew 10:28-31)

CHAPTER 10

STUDY QUESTIONS

1. Crabs, lobsters, eels, and hagfish.

2. a. God has built sharks out of cartilage, whereas the endoskeleton of fish is made of bone.
   b. Sharks have no covering for their gills.
   c. Sharks do not have the buoyancy bladder (bag).

3. a. Siamese fighting fish
   b. Whale shark
   c. Greenland shark
   d. Great white shark
   e. Irukandji box jellyfish
   f. Hagfish

4. Scientists believe that the salmon picks up on variations in magnetic forces along the earth's surface.

5. The female produces an egg, and the male produces a gametes.

SCRIPTURE

1. The LORD is my light and my salvation;
   Whom shall I **fear**?
   The LORD is the strength of my life;
   Of **whom shall I be afraid**?
   When the wicked came against me
   To **eat up my flesh**,
   My enemies and foes,
   They **stumbled and fell**.
   (Psalm 27:1-2)

2. And do not fear **those who kill the body** but cannot kill the soul. But rather fear Him who is

CHAPTER 11

STUDY QUESTIONS

1. 153

2. 2,500 pounds

3. 10.6 billion pounds. People eat 4 times more chicken than fish

4. Grass carp. In America, tuna and salmon are the most popular fish.

5. Alaska, Hawaii, and Japan

6. Mediterranean Sea, South America

SCRIPTURE

1. What is **man** that You are mindful of him,
   And the son of man that You visit him?
   For You have made him a little lower than the **angels**,
   And You have crowned him with glory and honor.
   You have made him to have **dominion** over the works of Your hands;
   You have put **all things under his feet**,
   All sheep and oxen—
   Even the beasts of the field,
   The birds of the air,
   And the **fish of the sea**
   That pass through the paths of the seas.
   (Psalm 8:4-8)

2. Then Jesus said to them, "Children, have you any food?" They answered Him, "No." And He said to

them, "**Cast the net** on the right side of the boat, and you will find some." So they cast, and now they were not able to draw it in because of the **multitude of fish**. Therefore that disciple whom Jesus loved said to Peter, "It is the Lord!" Now when Simon Peter heard that it was the Lord, he put on his outer garment (for he had removed it), and plunged into the sea. But the other disciples came in the little boat (for they were not far from land, but about two hundred cubits), dragging the net with fish. Then, as soon as they had come to land, they saw a fire of coals there, and **fish laid on it**, and bread. Jesus said to them, "Bring some of the fish which you have just caught." (John 21:5-10)

3. There will be a **very great multitude of fish**, because these waters go there; for they will be healed, and everything will live wherever the river goes. It shall be that **fishermen will stand by it** from En Gedi to En Eglaim; they will be places for spreading their nets. Their fish will be of the same kinds as the **fish of the Great Sea**, **exceedingly many**. (Ezekiel 47:9-10)

CHAPTER 12

STUDY QUESTIONS

1. a. Vertebrate
   b. Invertebrate, Arthropod, Insect, Exoskeleton
   c. Vertebrate, Amphibian
   d. Invertebrate, Crustacean, Exoskeleton
   e. Vertebrate, Cartilage, Endoskeleton
   f. Invertebrate, Arthropod, Insect, Exoskeleton
   g. Invertebrate
   h. Vertebrate, Scavenger
   i. Invertebrate, Arthropod, Insect, Metamorphosis, Exoskeleton
   j. Invertebrate, Arthropod, Arachnid, Exoskeleton
   k. Invertebrate, Mollusk, Gastropod, Nocturnal
   l. Invertebrate, Cephalopod
   m. Invertebrate, Arthropod, Arachnid
   n. Invertebrate, Arthropod, Exoskeleton

2. The tadpole has gills, a fish-like two-chambered heart, and a tail. The frog has lungs, a three-chambered heart, and two pairs of legs. Also, the mouth and digestive system will be different in the frog.

3. The mother frog lays eggs.
   Eggs hatch into tadpoles.
   Tadpoles turns into a frog.
   Frogs lay eggs.
   Frogs die.

4. Lateral lines are hair-like sensors which can sense pressure changes and motion in the water or air.

5. The saliva becomes less sticky, so the frog can swallow the insect.

6. a. Poison frogs
   b. Bull frog
   c. Darwin frog
   d. Glass frog
   e. Golden poison frog
   f. African clawed frog
   g. Salamander
   h. Wood frog

SCRIPTURE

1. [The Lord] has made the earth **by His power**,
   He has established the world **by His wisdom**,
   And has stretched out the heavens at His discretion.
   When He **utters His voice**,
   There is a multitude of waters in the heavens:
   And He causes the vapors to ascend from the ends of the earth.
   He makes lightning for the rain,
   He brings the wind out **of His treasuries**.
   (Jeremiah 10:12-13)

2. And the LORD spoke to Moses, "Go to Pharaoh and say to him, 'Thus says the LORD: "Let My people go, that they may serve Me. But if you refuse to let them go, behold, I will smite all your territory with **frogs**. So the river shall **bring forth frogs abundantly**, which shall go up and come into your house, into

GOD MADE ANIMALS - WORKBOOK

your **bedroom**, on your **bed**, into the houses of your servants, on your people, into your ovens, and into your **kneading bowls**. And the frogs shall come up on you, on your people, and on all your servants."'" (Exodus 8:1-3)

3. For behold,
He who **forms mountains**,
And **creates the wind**,
Who declares to man what his thought is,
And makes the morning darkness,
Who **treads the high places of the earth**—
The **LORD God of hosts** is His name.
(Amos 4:13)

CHAPTER 13

STUDY QUESTIONS

1. Snakes, dogs, and crocodiles.

2. A serpent.

3. The serpent would bite the heel of the Seed of the Woman. Jesus would crush the head of the serpent.

4. All animals ate plants and vegetables.

5. 121-131 ft (37-40 m). Blue whales can grow up to 100 feet long.

6. The tail of this animal is large and moves about like a cedar tree. He is first in the ways of God, which means that this is the most powerful and largest animal created by God. He eats grass like an ox, and some dinosaurs were certainly vegetarian.

7. A reptile is a cold-blooded animal. A bird is a warm-blooded animal. A reptile cannot fly like a bird. Most reptiles have four limbs, while birds have two legs.

8. A reptile does not have gills.

9. Anti-venom is the best treatment for snake bites.

10. a. Belcher's sea snake
    b. Crocodile
    c. Mary River turtle
    d. Reticulated python
    e. Pit viper
    f. Tuatara
    g. Saw scaled viper
    h. Giant tortoise
    i. King cobra

11. a. 30 g's.
    b. 18 ft
    c. 90 psi
    d. 12 inches
    e. 100,000 people

SCRIPTURE

1. So the LORD God said to **the serpent**:
"Because you have done this,
You are cursed more than all cattle,
And more than every beast of the field;
On your belly you shall go,
And you shall eat dust
All the days of your life.
And I will put enmity
Between you and the woman,
And between your seed and her Seed;
He shall **bruise your head**,
And you shall **bruise His heel**." (Genesis 3:14-15)

2. There are three things which are **too wonderful for me**,
Yes, four which I do not understand:
The way of an **eagle in the air**,
The way of a **serpent on a rock**,
The way of a ship in the midst of the sea,
And the way of a man with a virgin. (Proverbs 30:18-19)

3. And He said to them, "I saw **Satan** fall like lightning from heaven. Behold, I give you the authority to **trample on serpents and scorpions**, and over all the power of the enemy, and **nothing shall by any means hurt you**. Nevertheless do not rejoice in this, that the spirits are subject to you, but rather rejoice because your names are written in heaven." (Luke 10:18-20)

ANSWER KEY

CHAPTER 14

STUDY QUESTIONS

1. a. Indonesia
   b. Galapagos Islands
   c. Southwestern United States
   d. Madagascar (and Africa)
   e. Australia and Southeast Asia

2. Tremendous eyesight, a long tongue, the ability to mimic a leaf, sticky feet to climb up walls and over ceilings.

3. Lizards, salamanders, planarian flatworms, starfish

4. Both can shrink in size when there isn't enough food around.

5. a. Gila monster
   b. Komodo dragon
   c. Marine iguana
   d. Chameleon
   e. Crocodile
   f. Megalodon

6. Its head and eyes.

7. The crocodile's teeth stick out when its mouth is closed because its lower jaw is wider than its upper jaw. Also, the crocodile's snout tends to be narrower while the alligator's is more snub-snouted.

8. Bacteria.

SCRIPTURE

1. He stretches out the north over empty space;
   He hangs the earth on nothing.
   He binds up the water in His thick clouds,
   Yet the clouds are not broken under it.
   He covers the face of His throne,
   And spreads His cloud over it.
   He drew a circular horizon on the face of the waters,
   At the boundary of light and darkness.
   The pillars of heaven tremble,
   And are astonished at His rebuke.
   He **stirs up the sea** with His power,
   And by His understanding **He breaks up the storm**.
   By His Spirit He adorned the heavens;
   His hand **pierced the fleeing serpent**.
   Indeed these are the mere edges of His ways,
   And how small a whisper we hear of Him!
   But **the thunder of His power** who can understand?
   (Job 26:7-14)

2. Benaiah was the son of Jehoiada, the **son of a valiant man from Kabzeel**, who had done many deeds. He had killed two lion-like heroes of Moab. He also had gone down and **killed a lion in the midst of a pit** on a snowy day. (2 Samuel 23:20)

3. O LORD, our Lord,
   How **excellent is Your name in all the earth**,
   Who have set **Your glory above the heavens**!
   (Psalm 8:1)

CHAPTER 15

STUDY QUESTIONS

1. a. Vertebrate animals with mammary glands that nurse their babies. They have hair, they are warm-blooded, and usually give birth to fully developed babies.
   b. Vertebrate animals that are cold-blooded, created for both land and water. They go through metamorphosis.
   c. Vertebrate animals with scales. They are cold-blooded, they have a heart and lungs, and they generally have four limbs (except for snakes).
   d. Vertebrate animals equipped with wings. Birds have a heart and lungs, and they are warm-blooded.
   e. Vertebrate animals with a skeleton on the outside of their bodies — beetles, insects etc.

2. Answers:

a. Pouched Mammals	opossums, kangaroos

b. Placenta Mammals	dogs, horses, deer, whales
c. Egg-laying Mammals	duck-billed platypi, spiny anteaters
d. Insect-eating Mammals	moles, shrews
e. Flesh eating Mammals	dogs, cats, lions, bears
f. Gnawing Mammals	mice, squirrels, other rodents
g. Sea Mammals	whales, porpoises
h. Winged Mammals	bats
i. Proboscis Mammals	elephants
j. Even-toed Mammals	cattle, deer, pigs
k. Odd-toed Mammals	horses, rhinoceroses
l. Semi-erect Mammals	lemurs, apes, monkeys

3. a. Rats will control temperature using the blood vessels in their tails.
b. Dogs will cool themselves off by panting and blood vessels in their mouths cool off their blood. They will also use their body hair to keep warm in cold weather.
c. Humans will sweat to cool off.

4. Rats, dolphins, and pigs.

5. Dogs, parrots, pigs, and rats.

6. The Neanderthals had human DNA. They owned jewelry. They set broken bones. They conducted funerals. Humans do these things.

7. a. Three-banded armadillo -- rolls up in a perfect ball
b. Dormouse -- gives up his tail
c. Porcupine -- slaps quills from his tail
d. Hedgehog -- rolls himself into spiny ball and coats his spines with gross stuff
e. Opossum -- plays dead

8. Cockroaches -- walks on two feet for short distances
Parrots -- talks by mimicry
Crows -- makes tools to use
Beavers -- fashions its own house with rooms and openings

9. a. Majestic Animal -- Lion, Elk
b. Largest Land Animal -- Elephant
c. Dangerous Mammal -- Hippo, elephant, lion, cape buffalo
d. Unusual Mammal -- Duckbilled Platypus
e. Beautiful Animals -- Swan, white tiger, husky dog, peacock, panda bear
f. Funny-looking Animals -- Platypus and blobfish
g. Cutest Animals -- Wombat, slow loris, bottlenose dolphin, penguin, kittens, puppies, and arctic fox

10. a. Placenta -- The organ that provides food and oxygen to the baby in its mother's womb.
b. Belly button -- The mark left when the umbilical cord is removed after birth.
c. Marsupial -- A mammal that is equipped with a pouch for the baby to live in during most of its gestation.
d. Nocturnal -- Animals that are most active at night.
e. Diurnal -- Animals that are most active during the day.
f. Archeologists -- Scientists that look for evidences of ancient life and societies of humans and animals.

SCRIPTURE

1. Save me from the **lion's mouth**
And from the **horns of the wild oxen**! (Psalm 22:21)

2. There are three things which are **majestic in pace**,
Yes, four which are **stately in walk**.
A lion, which is mighty among beasts,
And does not turn away from any. (Proverbs 30:29-30)

3. All flesh is not the **same flesh**, but there is one kind of flesh of men, another flesh of **animals**, another of **fish**, and another of **birds**. (1 Corinthians 15:39)

4. One generation shall praise Your works to another,

ANSWER KEY

And shall declare Your **mighty acts**.
I will meditate on the glorious splendor of Your majesty,
And on Your **wondrous works**.
Men shall speak of the might of Your **awesome acts**,
And I will declare Your greatness.
They shall utter the **memory of Your great goodness**,
And shall sing of Your righteousness.
The LORD is gracious and full of compassion,
Slow to anger and great in mercy.
The LORD is good to all,
And His tender mercies are over **all His works**.
All Your works shall praise You, O LORD,
And Your saints shall bless You.
They shall speak of the glory of Your kingdom,
And talk of Your power. . . (Psalm 145:4-11)

CHAPTER 16

STUDY QUESTIONS

1. Bats and pit vipers

2. The Germans submarines were sinking many ships during World War I.

3. Fruit bats distribute seeds for tropical trees. They can also pollinate fruit plants.

4. Chicken and beef. Chickens give us 100 million metric tons of meat. Cattle give us 59 million tons of beef per year.

5. 1.48 billion acres.

6. a. Cows produce beef, a popular meat. They also produce milk. But, they are large and can be hard to control.
   b. Sheep are good grazers. Sheep milk is of higher quality than cow milk. Sheep produce wool and good meat. But, sheep can wander, and they are easily susceptible to attack from wild animals.
   c. Goats are very good at grazing, even mountain grazing. They are friendlier than sheep. They can make for good pack animals. Goat milk can be more digestible for children. They are not picky eaters. But, goats are escape artists and they can eat up a garden quite quickly. They are also susceptible to worms, and they will need regular nail clipping.
   d. Pigs are good at breaking up soil and clearing out property. They are pretty smart and they can figure out how to get out of a fenced area. But, pigs smell bad. They can grow pretty big (which is good for meat), but they can also hurt you.

7. a. Transportation and pulling logs (and heavy loads)
   b. Eats ants
   c. Transportation -- carries passengers and heavy loads
   d. Carries messages
   e. Meat for hunters
   f. Kills cobras/poisonous snakes
   g. Eats dangerous spiders, harmful bugs
   h. Angora wool for sweaters, etc.

8. a. Bad Dominion
   b. Bad Dominion
   c. Good Dominion
   d. Good Dominion
   e. Bad Dominion

9. a. Siberian husky
   b. Sheep dogs
   c. The retrievers, the beagle, the setters, the English springer spaniel, and the English pointer.
   d. The Akita dog, the German shepherd, the giant schnauzer, the Doberman pinscher, the bull mastiff, and the Caucasian shepherd dog.

SCRIPTURE

1. Then the LORD sent Nathan to David. And he came to him, and said to him: "There were two men in one city, one rich and the other poor. The rich man had exceedingly **many flocks and herds**. But the poor man had nothing, except one **little ewe lamb** which he had bought and **nourished**; and it grew up together with him and with his children. It ate of his own food and drank from his own cup and **lay in his bosom**; and it was like **a daughter to him**. (2 Samuel 12:1-3)

2. Now while he was still speaking with them, Rachel came with **her father's sheep**, for she was **a shepherdess**. And it came to pass, when Jacob saw Rachel the daughter of Laban his mother's brother, and the sheep of Laban his mother's brother, that Jacob went near and rolled the stone from the well's mouth, and **watered the flock of Laban** his mother's brother. (Genesis 29:9-10)

3. If an ox **gores a man or a woman to death**, then the ox shall surely be stoned, and its flesh shall not be eaten; but the owner of the ox shall be acquitted. But if the ox tended to **thrust with its horn in times past**, and it has been made known to his owner, and he has not kept it confined, so that **it has killed a man or a woman**, the ox shall be **stoned** and its owner also shall **be put to death**. (Exodus 21:28-29)

4. Be diligent to know **the state of your flocks**,
And attend to your **herds**;
For riches are not forever,
Nor does a crown endure to all generations.
When the hay is removed, and the tender grass shows itself,
And the herbs of the mountains are gathered in,
The lambs will **provide your clothing**,
And the goats **the price of a field**;
You shall have enough **goats' milk for your food**,
For the **food of your household**,
And the nourishment of your maidservants.
(Proverbs 27:23-27)

FINAL EXAMINATION

A. TRUE OR FALSE

1. True
2. True
3. False
4. False
5. True
6. True
7. False
8. True
9. True
10. False
11. True
12. False
13. True
14. True
15. False

B. DEFENSE MECHANISMS

16. Plays dead
17. Hides among the sea anemone
18. Shoots out poison though a spiny protrusion
19. Rolls itself into a perfect little ball
20. Changes colors to blend in with environment
21. Licks poison onto its elbows

C. UNIQUE GIFTS

22. Sonar
23. Infrared detection
24. Mammary glands
25. Lateral lines

D. SHORT ANSWER

26. Genesis 1 and Psalm 8

ANSWER KEY

27. Anti-venom

28. Tissue

29. Invertebrates -- Jellyfish, Mollusks, Worms, Starfish, etc.

30. Scavengers -- Vultures, etc.